# Lecture Notes in Computer Science 7404

Commenced Publication in 1973
Founding and Former Series Editors:
Gerhard Goos, Juris Hartmanis, and Jan van Leeuwen

Gonzalo Navarro   Vladimir Pestov (Eds.)

# Similarity Search and Applications

5th International Conference, SISAP 2012
Toronto, ON, Canada, August 9-10, 2012
Proceedings

 Springer

Volume Editors

Gonzalo Navarro
Universidad de Chile
Departamento de Ciencias de la Computación
Blanco Encalada 2120, Santiago, Chile
E-mail: gnavarro@dcc.uchile.cl

Vladimir Pestov
University of Ottawa
Department of Mathematics and Statistics
585 King Edward Avenue, Ottawa, ON, Canada, K1N 6N5
E-mail: vpest283@uottawa.ca

ISSN 0302-9743               e-ISSN 1611-3349
ISBN 978-3-642-32152-8       e-ISBN 978-3-642-32153-5
DOI 10.1007/978-3-642-32153-5
Springer Heidelberg Dordrecht London New York

Library of Congress Control Number: 2012942592

CR Subject Classification (1998): H.3.1, I.5.3, E.1, H.3.3, H.2.4, H.2.8, F.2.2, G.1.2-3

LNCS Sublibrary: SL 3 – Information Systems and Application, incl. Internet/Web and HCI

*Typesetting:* Camera-ready by author, data conversion by Scientific Publishing Services, Chennai, India

Printed on acid-free paper

Springer is part of Springer Science+Business Media (www.springer.com)

# Preface

This volume contains the papers presented at the 5th International Conference on Similarity Search and Applications (SISAP 2012), which took place during August 9–10, 2012, at the Fields Institute for Research in Mathematical Sciences, Toronto, Ontario, Canada.

SISAP is a conference devoted to similarity searching, with emphasis on metric space searching. It aims to fill in the gap left by the various scientific venues devoted to similarity searching in spaces with coordinates, by providing a common forum for theoreticians and practitioners around the problem of similarity searching in general spaces (metric and non-metric) or using distance-based (as opposed to coordinate-based) techniques in general. Four types of contributions are welcome: (1) fundamental techniques to handle general similarity search problems, (2) applied techniques to solve particular similarity search problems of wide interest, (3) new similarity search problems, where their features and challenges are studied, and (4) actual systems for similarity search, in the form of demos. SISAP is seen as a forum for not only exchanging new indexing techniques and real-world applications, but also common testbeds and benchmarks, and source code. Authors are expected to use the testbeds and code from the SISAP website (www.sisap.org) for comparing new applications, databases, indexes, and algorithms.

This year we received 19 full-paper and two demo submissions, from Argentina, Chile, Czech Republic, France, Japan, Mexico, Norway, Russia, Spain, Switzerland, UK, and USA. Each submission was assigned, in double-blind mode, to three Program Committee (PC) members, who reviewed them themselves and/or supervised subreviews. Submissions received two to five reviews (3.14 on average). Then the PC Chairs and involved members discussed the articles where no obvious agreement had been reached. The final decisions of acceptance or rejection were made by the PC Chairs. Finally, 14 full papers and the two demos were selected to be presented at the conference and to appear in the proceedings.

Of the full papers accepted, nine refer to techniques to handle general similarity search problems, improving upon the state of the art on topics like parallelism, dynamism, secondary memory, approximation techniques, optimized construction, combinations of data structures, and novel scenarios, such as streams of related searches and inferring factual space properties from the data. Further, two accepted papers refer to applied techniques, to similarity searching in string dictionaries and in images. The other three papers study the properties of specific spaces such as sequences under time-warping distance and factorized tensors, and propose and study new distances for vector spaces based on entropy correlations. Of the two demos, one presents an image meta-search engine, and the other introduces a tool for identifying protein and peptide sequences from tandem mass spectra.

Overall, the articles formed an extremely stimulating set of contributions to many of the most relevant aspects of similarity searching. Two invited presentations and papers from prominent researchers further enriched this year's SISAP. The first one, "Effective Principal Component Analysis," by Santosh Vempala, is about the success and challenges around this technique, of wide relevance in similarity search and various other fields. The second one, "Future Trends in Similarity Searching," by Pavel Zezula, is a revealing survey and analysis of where the discipline is expected to head in the forthcoming years.

This year the proceedings of SISAP were published by Springer-Verlag, in the *Lecture Notes in Computer Science* series. A selection of the best papers was recommended for inclusion in a special issue of the *Information Systems* journal dedicated to this conference. These were chosen by the PC Chairs based on the original reviews of the articles and their oral presentation during the conference, as well as appropriateness to the journal.

The subject matter of the SISAP conferences, although primarily a computer science topic, uses a great deal of advanced mathematical methods, such as those of geometric functional analysis and statistical machine learning. The conference is a perfect platform for interactions between computer scientists and mathematicians, and the stimulating research ambiance of the Fields Institute gave fresh impetus to such interactions. We thank the Fields Institute for the hosting of SISAP 2012 conference.

Last, but not least, we acknowledge the generous financial support from (again) the Fields Institute for Research in Mathematical Sciences, Canada; the Canadian Network of Excellence in Mathematics of Information Technology and Complex Systems (MITACS); and the Natural Sciences and Engineering Research Council of Canada (NSERC) research grant "New Set-Theoretic Tools for Statistical Learning." All the submission, reviewing, and proceedings generation processes were handled through the EasyChair platform.


August 2012                                              Gonzalo Navarro
                                                        Vladimir Pestov

# Organization

## Committees

### Steering Committee

| | |
|---|---|
| Edgar Chávez | Universidad Michoacana, Mexico |
| Gonzalo Navarro | Universidad de Chile, Chile |

### Program Committee Chairs

| | |
|---|---|
| Gonzalo Navarro | Universidad de Chile, Chile |
| Vladimir Pestov | Université d'Ottawa, Canada |

### Program Committee Members

| | |
|---|---|
| Edgar Chávez | Universidad Michoacana, Mexico |
| Paolo Ciaccia | Università di Bologna, Italy |
| Alfredo Ferro | Università di Catania, Italy |
| Daniel Keim | Universität Konstanz, Germany |
| Daniel Miranker | University of Texas at Austin, USA |
| Marco Patella | Università di Bologna, Italy |
| Hanan Samet | University of Maryland, USA |
| Tomáš Skopal | Charles University in Prague, Czech Republic |
| Aleksandar Stojmirović | NCBI/NLM/NIH, USA |
| Agma Traina | Universidade de São Paulo – São Carlos, Brazil |
| Pavel Zezula | Masaryk University, Czech Republic |

### Organization Chair

| | |
|---|---|
| Vladimir Pestov | Université d'Ottawa, Canada |

### Publicity Chair

| | |
|---|---|
| Tomáš Skopal | Charles University in Prague, Czech Republic |

## Additional Reviewers

| | | |
|---|---|---|
| Marco Adelfio | David Hoksza | Luís M. Silveira Russo |
| Gelio Alves | Eamonn Keogh | Jan Sedmidubsky |
| Michal Batko | Jakub Lokoč | John Spouge |
| Petra Budikova | Jiří Novák | Eric Sadit Téllez Avila |
| Benjamin Bustos | Sarana Nutanong | Lee Thompson |
| Carlos Castillo | Ives Pola | German Tischler |
| Vlastislav Dohnal | Mônica R. Porto Ferreira | Kesheng Wu |
| Magnus Lie Hetland | Nora Reyes | |

## Sponsoring Institutions

Fields Institute for Research in Mathematical Sciences, Canada
Canadian Network of Excellence in Mathematics of Information Technology and
Complex Systems (MITACS)
Natural Sciences and Engineering Research Council of Canada (NSERC)

# Table of Contents

## Invited Papers

## New Scenarios and Approaches

## Improving Metric Data Structures

## Facing Scalability Issues

## Searching in Specific Spaces

## New Similarity Spaces

## Demo Papers

# Effective Principal Component Analysis

Santosh S. Vempala*

School of CS and Algorithms and Randomness Center,
Georgia Tech
Atlanta, GA, USA
vempala@gatech.edu

**Abstract.** Principal Component Analysis (PCA) is one of the most widely used algorithmic techniques. When is PCA provably effective? What are its main limitations and how can we get around them? In this note, we discuss three specific challenges.

## 1 Introduction

Algorithms for processing large, high-dimensional data sets are of increasing importance. The field of sublinear algorithms is motivated by the abundance of data and the practical infeasibility of running algorithms even with (large) polynomial time complexity on such inputs.

In this endeavor, some classical approaches have flourished. Notable among them is Principal Component Analysis, a widely-used, general purpose method for summarizing data, compactly representing it, clustering, finding near(est)-neighbors, and many other applications that can be classified as finding patterns or data mining. It serves as a ubiquitous pre-processing step.

What is PCA? Given an $m \times n$ real matrix $A$, with rows interpreted as points and columns as coordinates, standard PCA attempts to find the most important combinations of coordinates, i.e., the subspace that best approximates the data upon projection. More precisely, we first shift the matrix so that the average of the rows is zero, then the top principal component is the unit vector $v^1$ which maximizes $\|Av\|^2$, i.e.,

$$v^1 = \arg\max_{v:\|v\|=1} \|Av\|.$$

Similarly the top $k$ principal components span a subspace $V$ of dimension $k$ that maximizes the norm of the projection of the points to it. In particular, if the data were inherently $k$-dimensional, then PCA would preserve the points and recover the entire norm of the original matrix.

PCA is usually implemented by the Singular Value Decomposition (SVD) applied to a centered (mean zero) matrix. For an $m \times n$ real matrix $A$, the SVD is defined as follows:

$$A = UDV^T$$

---

* This work was partially supported by the NSF.

G. Navarro and V. Pestov (Eds.): SISAP 2012, LNCS 7404, pp. 1–7, 2012.
© Springer-Verlag Berlin Heidelberg 2012

where $U \in \mathbb{R}^{m \times r}$ and $V \in \mathbb{R}^{n \times r}$ are column-orthonormal matrices, with the columns $u^i$ of $U$ being *left singular vectors* and the columns $v^i$ of $V$ being *right singular vectors*; the matrix $D = \text{diag}(\sigma_1, \ldots, \sigma_r)$ is a positive diagonal matrix and by convention, the diagonal entries in $D$, the *singular values* $\sigma_i$, are non-increasing. The top $k$ principal components of $A$ as defined above are simply the first $k$ columns of $V$. The top left and right singular vectors $u^1, v^1$, the first columns of $U$ and $V$ respectively, are given by

$$(u^1, v^1) = \arg \max_{\|u\|=\|v\|=1} u^T A v.$$

The next such pair maximizes the same function among vectors orthogonal to $u^1$ and $v^1$.

PCA/SVD readily give a rank-$k$ representation of $A$, namely,

$$A_k = \sum_{i=k} \sigma_i u^i (v^i)^T = A V_k V_k^T$$

where $V_k$ is the submatrix of the first $k$ columns of $V$. This representation gives a least-squares approximation as shown by Eckart and Young [11].

**Theorem 1.** *[11] For any $m \times n$ matrix $A$, and any integer $1 \leq k \leq n$, among all matrices of rank at most $k$, we have*

$$A_k = \arg \min_{D: rank(D) \leq k} \|A - D\|_F = \arg \min_{D: rank(D) \leq k} \|A - D\|_2.$$

Thus, PCA can be used to obtain a low-dimensional representation of data, an important pre-processing step for many applications. There is considerable work on making the PCA algorithm efficient, both in numerical linear algebra where the focus is more on the convergence rate of algorithms as a function of the error bound, and in algorithmic complexity theory, where the focus has been to find algorithms with bounded error whose complexity is sublinear in the size of the matrix. For further reading, the reader is referred to [16].

## 2  One Success and Three Challenges

The impact of PCA in practice is hard to exaggerate. In many empirical studies, PCA or PCA-based algorithms match or outperform other methods. In spite of a plethora of such results (many quite striking), it is not hard to see that PCA can fail. For example, if one is interested in reducing dimensionality to make nearest neighbor search more efficient, then PCA can easily destroy the neighborhood structure of a subset of the data. Similarly, if one is interesting in recovering a clustering of an unlabeled data set, even a set with a clustering defined by a simple rule (e.g., a halfspace), PCA can collapse the clusters together.

In light of these examples, it is natural to wonder when PCA is provably effective and what aspects of a problem or data set can render it ineffective. While we are far from a satisfactory answer to these questions, in this note we illustrate ongoing research in this direction and highlight three challenges for PCA via one important problem. The challenges are: (1) PCA is not affine-invariant (2) PCA takes into account only pairwise relationships (matrices, second moments) and thus is a crude approximation for tensor data (multidimensional arrays, higher moments) and (3) PCA is not noise-tolerant.

The problem is unsupervised clustering. Many special cases of this problem are widely studied, including (1) unraveling a mixture of Gaussians given i.i.d. samples from the mixture and (2) detecting a planted clique in a random graph. We note that both problems fit under the framework of inferring a mixture model from samples: in one case the components of the mixture are Gaussians, while in the other case the components are distributions on binary hypercube.

One approach to understanding high-dimensional data is to model it as a distribution that is itself a mixture (a convex combination) of a small number of "simple" distributions. The most common variant of such a mixture model is to assume that the component distributions are unknown Gaussians. The problem is thus to find the component Gaussians (their means and covariances) and the mixing weights from samples. There is a large literature on this problem and there has been much progress on understanding its complexity over the past 15 years [9, 2, 10, 19, 15, 1, 7, 6, 5, 14, 4, 18, 3].

The current state of the art is that under some assumptions on the probabilistic separability of the components, the mixture can be learned in time polynomial in both $n$ and $k$ [6]. Alternatively, under no separability assumptions, a mixture of $k$ Gaussians can be learned in time exponential in $k$ [14, 4, 18, 3].

An important step in the line of work on polynomial-time algorithms is PCA. In [19], the following algorithm to learn a $k$-component mixture is analyzed.

---

**Basic Spectral Algorithm**

1. Apply PCA to project data to a $k$-dimensional subspace.
2. Partition samples into $k$ components using only their pairwise Euclidean distance in the $k$-dimensional space.
3. Using the samples of each cluster, estimate the mean and covariance of each component.

---

The following guarantee holds for this algorithm. Here $w_i$ are the mixing weights, with $w_{min}$ being the smallest, and the $i$'th component is $N(\mu_i, \sigma_i^2 I_n)$.

**Theorem 2.** *[19] Let F be a mixture of k spherical Gaussians with the property that for any pair of components $i, j$,*

$$\|\mu_i - \mu_j\| \geq C \max\{\sigma_i, \sigma_j\}(k \log(n/w_{min}))^{1/4}.$$

*Then given at least $poly(n, 1/w_{min})$ i.i.d. samples from the mixture, with high probability the sample is correctly classified by the Basic Spectral Algorithm.*

(for more explicit bounds on the number of samples and the failure probability, we refer to [19]).

This result remains the best known for unraveling a mixture of spherical Gaussians. Information-theoretically, a separation that grows with $\log k$ rather than poly($k$) is possible, but no efficient method is known, and PCA does not work. When we consider the algorithm for a general (nonspherical) Gaussian mixture, this failure is magnified. For PCA to work, one has to assume that the component means are separated by a distance that grows with both $\sqrt{k}$ and the *largest standard deviation* of the components. To see this, if one considers a mixture of two Gaussians that are spherical in $n - 1$ directions and very thin in one direction arranged to look like two parallel "pancakes", then unless the two Gaussians are separated by a distance proportional to the larger standard deviation, PCA will collapse them.

This brings us to our first challenge, namely PCA is not affine-invariant. The separability of two Gaussians is an affine-invariant property (if two Gaussians have small overlap, any affine transformation of them has the same overlap). For the two pancake example, if one were to stretch space along the thin direction of the pancakes, then a significantly smaller separation suffices.

At first glance, an affine-invariant version of PCA sounds self-contradictory. After all, PCA is designed to pick up the most significant directions of a data set, as measured by the second moment. What if all second moments are equal, i.e., the distribution is isotropic? An immediate possibility is to consider higher moments. Namely, find directions that maximize higher moments. This natural generalization runs into several difficulties. First, the maximization problem is NP-hard. Second, there can be an exponential number of local maxima and there is little structure to them unlike in the case of the second moment when they are pairwise orthogonal.

In spite of these difficulties, in [6], an affine-invariant method is given that can be viewed as an efficient algorithm for maximizing a certain linear combination of higher moments. It is used to unravel a mixture of arbitrary Gaussians assuming only that each Gaussian is separated from the rest of the mixture by a hyperplane (this separation condition is itself affine-invariant). It remains open to solve the problem under the weakest assumption of probabilistic separability. We describe the algorithm below and refer the reader to [6] for details on its guarantees.

---

**Isotropic PCA**

1. Apply an isotropic transformation to the input data, so that the mean of the resulting data is zero and its covariance matrix is the identity.
2. Give a weight to each point using the density of a spherically symmetric weight function centered at zero, e.g., a spherical Gaussian.
3. Perform PCA on the weighted data.

---

Another approach to dealing with this issue is the one taken by Independent Component Analysis (ICA) [13], where data is assumed to be generated by applying an arbitrary affine transformation to a product distribution, and the goal is to recover the basis of the underlying product distribution. It is shown in the literature that finding *local optima* of higher moments allows one to solve this problem under rather mild assumptions (specifically that the one-dimensional distributions are not too close to Gaussian) [8, 17, 12]. In recent work [20], this was extended to factoring a distribution that is a product of its marginals on orthogonal subspaces (not necessarily one-dimensional). The approach is to find directions that maximize higher moments. We note that the direction that maximizes the $r$'th moment is equivalent to the vector (principal component) that achieves the norm of an $r$-dimensional array defined by the expectations of products of $k$-subsets of coordinates, i.e., local maxima of the function

$$\sum_{i_1,\ldots,i_r} A_{i_1,\ldots,i_r} v_{i_1}^1 v_{i_2}^2 \ldots v_{i_r}^r.$$

In the case of ICA, the entries of $A$ are defined as $A_{i_1,\ldots,i_r} = \mathsf{E}(x_1 x_2 \ldots x_r)$. While this approach has already lead to some interesting results already, it has the promise of wider applicability and has not been thoroughly understood.

Our final challenge is noise-tolerance. PCA is rather brittle in that adding a single point sufficiently far from the mean can change the top principal component arbitrarily (and adding $k$ points can similarly affect the top $k$ pirncipal components). So if the mixture is modified by a small amount of arbitrary noise then PCA could give very different results. How can we address this? In the context of Gaussian mixtures, or unsupervised clustering, it is very natural (and probably essential in any practical setting) that there is some amount of noise that does not fit the model. All known algorithms for mixture models are unable to handle noise, with one exception.

The exception is Brubaker's noise tolerant unsupervised clustering algorithm [5]. He was able to show that his algorithm works for a mixture of separable logconcave component distributions (significantly more general than Gaussian components), with inter-mean separation only a logarithmic factor higher than the known bound for PCA for the perfect mixture setting with no noise. Under this stronger separation, the algorithm works for a small amount of arbitrary noise. It is not known whether this method can also be used to make other algorithms agnostic, e.g., the current best results for spherical Gaussian mixtures (Theorem 2) or the affine-invariant separation results, or the exponential algorithms that make no separability assumptions.

The most interesting part of the algorithm is the robust variant of PCA introduced in Brubaker's paper.

---

**Robust PCA**

1. Given distribution $F$ in $R^n$, set current dimension $d = n$.
2. While $d > k$,
   (a) Make $F$ isotropic.
   (b) Restrict $F$ to points within a ball of radius $C\sqrt{d}$ (where $C$ is a fixed constant).
   (c) Project $F$ to span of its top $\lfloor (d - k)/2 \rfloor + k$ principal components.
3. Output current subspace of dimension $k$.

Thus the algorithm alternates between removing outliers and reducing the dimension. The important thing to note is that in both phases it is conservative, choosing only to remove extreme outliers and reducing dimension by discarding roughly the bottom half of the principal components. Brubaker showed this algorithm gives an agnostic clustering method for mixtures of logconcave distributions, but it remains open to fully understand this robust PCA algorithm. In particular, is the subspace output by the algorithm nearly invariant under small changes to the data set?

# References

[1] Achlioptas, D., McSherry, F.: On Spectral Learning of Mixtures of Distributions. In: Auer, P., Meir, R. (eds.) COLT 2005. LNCS (LNAI), vol. 3559, pp. 458–469. Springer, Heidelberg (2005)

[2] Arora, S., Kannan, R.: Learning mixtures of arbitrary gaussians. Annals of Applied Probability 15(1A), 69–92 (2005)

[3] Belkin, M., Sinha, K.: Polynomial learning of distribution families. In: FOCS, pp. 103–112 (2010)

[4] Belkin, M., Sinha, K.: Toward learning gaussian mixtures with arbitrary separation. In: COLT, pp. 407–419 (2010)

[5] Brubaker, S.C.: Robust pca and clustering on noisy mixtures. In: Proc. of SODA (2009)

[6] Brubaker, S.C., Vempala, S.: Isotropic pca and affine-invariant clustering. In: Grötschel, M., Katona, G. (eds.) Building Bridges Between Mathematics and Computer Science. Bolyai Society Mathematical Studies, vol. 19 (2008)

[7] Chaudhuri, K., Rao, S.: Learning mixtures of product distributions using correlations and independence. In: Proc. of COLT (2008)

[8] Comon, P.: Independent Component Analysis. In: Proc. Int. Sig. Proc. Workshop on Higher-Order Statistics, Chamrousse, France, July 10-12, pp. 111–120 (1991); Keynote address. Republished in Lacoume, J.L. (ed.): Higher-Order Statistics, pp 29–38. Elsevier (1992)

[9] DasGupta, S.: Learning mixtures of gaussians. In: Proc. of FOCS (1999)

[10] DasGupta, S., Schulman, L.: A two-round variant of em for gaussian mixtures. In: Proc. of UAI (2000)

[11] Eckart, C., Young, G.: The approximation of one matrix by another of lower rank. Psychometrika 1(3), 211–218 (1936)

[12] Frieze, A., Jerrum, M., Kannan, R.: Learning linear transformations. In: FOCS, pp. 359–368 (1996)

[13] Jutten, C., Herault, J.: Blind separation of sources, part i: An adaptive algorithm based on neuromimetic architecture. Signal Processing 24(1), 1–10 (1991)

[14] Kalai, A.T., Moitra, A., Valiant, G.: Efficiently learning mixtures of two gaussians. In: STOC, pp. 553–562 (2010)

[15] Kannan, R., Salmasian, H., Vempala, S.: The spectral method for general mixture models. SIAM Journal on Computing 38(3), 1141–1156 (2008)

[16] Kannan, R., Vempala, S.: Spectral algorithms. Foundations and Trends in Theoretical Computer Science 4(3-4), 157–288 (2009)

[17] Lacoume, J.-L., Ruiz, P.: Separation of independent sources from correlated inputs. IEEE Transactions on Signal Processing 40(12), 3074–3078 (1992)
[18] Moitra, A., Valiant, G.: Settling the polynomial learnability of mixtures of gaussians. In: FOCS, pp. 93–102 (2010)
[19] Vempala, S., Wang, G.: A spectral algorithm for learning mixtures of distributions. Journal of Computer and System Sciences 68(4), 841–860 (2004)
[20] Vempala, S., Xiao, Y.: Structure from local optima: Learning subspace juntas via higher order pca. CoRR, abs/1108.3329 (2011)

# Future Trends in Similarity Searching

Pavel Zezula

Masaryk University,
Botanicka 68a, Brno, Czech Republic
zezula@fi.muni.cz

**Abstract.** Similarity searching has been a research issue for many years, and searching has probably become the most important web application today. As the complexity of data objects grows, it is more and more difficult to reason about digital objects otherwise than through the similarity. In this article, we first discuss concepts of similarity and searching in light of future perspectives before a concise history of similarity searching technology is presented. We use the historical knowledge to extend the trends to future. We analyze the bottlenecks of application development and discuss perspectives of search computing for future applications. We also present a model of search technology and its position in computer clouds for application development. Finally, execution platforms for multi-modal findability and security issues for outsourced similarity searching environments are suggested as important research challenges.

## 1 Introduction

A wide range of data processing applications need to access data through similarity. Search is also one of the fundamental topics in computer science. Given a set of points from a universe and a distance measure, it is possible to pose similarity search queries on a sample point. It is an important operation in multimedia databases, information retrieval, biological databases, CAD parts in engineering environments, and many other database applications involving complex objects. Besides being used directly, it is also applied as a basic operation of more complex applications, such as data mining, decision making, event detection, or augmented reality. Similarity search no doubt already witnessed a great commercial success, but we believe that many others are to follow. However, the new possibilities are also posing new research challenges. Some of them are outlined in this paper.

In the following, we first discuss in Section 2 importance and future of similarity and search in our lives. In Section 3, we present a concise history of similarity searching technology and try to extend past trends to future. Suggested objectives of a better similarity search technology start in Section 4 with analyzing the state of the art and end with a definition of a generic model of similarity searching for cloud services. Finally, the problems of findability and security in such environments are discussed in Section 5 as important research challenges.

G. Navarro and V. Pestov (Eds.): SISAP 2012, LNCS 7404, pp. 8–24, 2012.

# 2   On the Importance of Similarity and Searching

## 2.1   Similarity

According to the WordNet[1] (a semantic lexicon of the English language):

> *Similar real world entities are marked by correspondence or resemblance, have the same or close characteristics, express closely related meanings, or are capable of replacing or changing places with something else. The noun "similarity" has one meaning, which is the quality of being similar.*

Because no situation, object, nor event is the same in all respects to any previous encountered situation, object, or event, using past experience to guide future behavior requires generalizing from previous to new instances. It is widely assumed, in *behaviorist* as well as *cognitive* theories of learning, that this generalization is based, to some degree at least, on similarity. Similarity appears also to have an important role to play in problem solving, inference and scientific reasoning, especially if analogy is viewed as a special case of similarity.

Similarity is a central notion throughout *cognitive sciences* [1]. In *perception*, the similarity between sets of visual or auditory stimuli influences the way in which they are grouped. In *speech* recognition, the similarity between different phonemes determines how easily confused they are. In *classification*, the category assignment to an instance may be influenced by the similarity of the new instance to past instances or a stored prototype. In *memory*, it has been suggested that retrieval from a cue depends on the similarity of past memory traces to the representation of the cue. Similarity also appears fundamental to learning and development.

However, not all application areas use the term similarity in exactly the same meaning. Consider the following examples. In the *geometry*, the similarity is seen as the relationship between two- or three-dimensional figures having the same shape but not necessarily the same size. The angles of two similar polygons or solids are equal, but the lengths of their sides are only proportional. Observe that given a specific polygon, any other possible polygon is either similar or it is not similar at all.

On the other hand, in order to assess the similarity between two molecules, say A and B, we need to first describe the molecules according to some scheme and then choose an appropriate measure to compare the descriptions of the molecules. A common method is to create a binary string where a bit or a set of bits being set on implies the presence of a particular feature. This process is independent of the measure we may use to associate or compare these binary strings, but in general, it is based on a *gradual comparison* of pairs of molecules. Given a molecule A and specific similarity measure, any other molecule B is in a relation with A, where the degree of similarity determines a ranking or closeness of B with respect to A. If we repeat this process for a large number of molecules,

---

[1] http://wordnet.princeton.edu/

we introduce another independent process of *classification, categorization,* or *clustering.*

Consider a simple and appealing idea about the way people decide whether an object belongs to a category, [2]: the object is a member of specific category if it is sufficiently similar to known category members. To put it in more cognitive terms, if you want to know whether an object is a category member, start with a representation of the object and a representation of the potential category. Then, determine the similarity value of the object representation to the category representation. If this similarity value is high enough, then the object belongs to the category; otherwise, it does not.

For example, suppose you come across a white three-dimensional object with an elliptic profile – or suppose you read or hear a description like this. You can calculate a measure of the similarity between your mental representation of this object and your prior representation of categories it might fit into. Depending on the outcome of this calculation, you might decide that similarity warrants calling the object an egg, perhaps, or a turnip, or a Christmas ornament.

In *social psychology*, similarity refers to how closely attitudes, values, interests, and personality match between people – the study of interpersonal attractions [3] is a major area of study in social psychology. Research has consistently shown that similarity leads to *interpersonal attraction*, that is the attraction between people which leads to friendships and romantic relationships. Many forms of similarity have been shown to increase liking. Similarities in opinions, interpersonal styles, amount of communication skills, demographics, and values have all been shown in experiments to increase liking. Of all the decisions people make that affect their environment, choosing friends and spouses are among the most important. For humans, both spouses and best friends are most similar on socio-demographic variables, such as age, ethnicity and educational level, next most on opinions and attitudes, then on cognitive ability, and least, but still significantly, on personality and physical traits. Having relationships with similar people helps to validate the values held in common. At the same time, people tend to make negative assumptions about those who disagree with them on fundamental issues, and hence feel repulsion.

The ability to *perceive similarities* is one of the most fundamental aspects of human cognition. Besides being crucial for recognition, classification, and learning and it plays an important role in scientific discovery and creativity. In recent years, similarity and analogy have received increasing attention from cognitive scientists. This growth of interest is related to the realization that human reasoning does not always operate on the basis of content-free general inference rules but, rather, is often tied to particular bodies of knowledge and is greatly influenced by the *context* in which it occurs. In a reasoning system of this kind, learning does not get accomplished by merely adding new facts and applying the same inference rules to them. Rather, successful learning often depends on the ability to identify the most relevant bodies of knowledge that already exist in memory so that this knowledge can be used as the starting point for learning something new.

Similarity may seem to be an irreducible *psychological primitive*, like the concept of "red" in the context of colors, but various theorists have tried to show how it relates to other fundamental considerations. For example, the representation of red varies widely across categories such as apple, brick, face, hair, light, soil, wine, and so forth. One could argue that red is stable only within particular categories rather than across all categories.

In summary, similarity has many faces even for the same object. It is subjective and depends on context, which is obviously changing in time. On the theoretical level, it is generally assumed that stimulus features are internally represented and that similarity between objects comes from some sort of comparison between their representations. Then, several interlocking questions may occur: What information is carried in the stimulus representations? How is this information combined or structured within the representation? How are representations compared in arriving at a "similarity"? Given a set of stimuli, how are their similarities determined and best represented?

## 2.2   Searching

*An activity of looking thoroughly in order to find something or someone, an investigation seeking answers, an operation that determines whether one or more of a set of items has a specified property, the examination of alternative hypotheses, and even the boarding and inspecting a ship on the high seas,*

are designated by WordNet as search. In the contemporary digital world, search tools are designed to help us *ask, browse, learn, share, visualize,* and *understand* the vast collection of sundry facts, nowadays available in digital form. As observed in [4], almost everything that we see, read, hear, write, measure, or otherwise observe can now be digital.

Research reports in management indicate that employees spend roughly 25 to 35 percent of their time searching for information they need to do their jobs. That's approximately one week of each month spent looking for data, which translates into billions of dollars in lost productivity time for employers. Search tools should help customers manage and utilize their existing business information – from e-mails and documents to scanned images, audio and video – to help them become more efficient, respond faster to suppliers and customers and tap into new markets, and make their operations more cost-effective.

According to [5], search is motivated by the need to *find objects* and *answers*. For example, we seek to find a web page with specific content, workshops able to repair our cars, holidays to relax, posters to decorate our rooms, or facts about earthquakes and tsunami. It is a process that leads from a query to desired results. Obviously, search is not the only way we find. We can ask professionals, friends, acquaintances, or colleges – family members are usually the first to ask. In any case, when we ask and search, our goal is to find. Our strategies for asking are often stimulated by place and time, that is the context, in general. Different questions are asked during university course exams, banquets with friends, and when finding our way to a museum in a foreign city.

But sometimes, we get an answer even without asking. This typically comes from once established relationships, memberships, location or identity, which can be seen as an ongoing query over universal data-set. We also *browse* to find, but it takes time and to find good things, you need a good luck. In fact, the process of finding is continuously and repeatedly changing the modes of *asking, browsing, filtering* and *searching* to achieve the goal. Finding strategies should consider all of these modes. We must aim for searcher's intent, we must respect what they want and need.

According to [6], search is not only about finding. Search at its best is about *conversation*. It is an interactive and iterative process where we find and learn. The answer changes the question. The process moves the goal. Search has the power to *suggest, define, refine, cross-sell, relate,* and *educate*. In fact, search has already registered enormous commercial success and it is among the most influential ways we learn. It is trusted and relied upon by millions of people a day. Searching on servers like Google, Yahoo, or Bing is the world's most popular teacher.

However, search is a difficult problem of terrific consequences, and it is not a solved problem. It changes the way we find everything from answerers, articles, and advertising to products, people, and places. It shapes how we learn and what we believe. It informs and influences our decisions. It thrives with and across myriad objectives and contexts, which is the promise for future applications.

According to recent studies [7], we are increasingly handing off the job of re-membering to search engines. Specifically, assuming information continually and instantaneously available on the web, our *cognitive habits* change. Experiments showed that when we do not know an answer to a question, our first reaction is to quickly find the nearest Web connection, rather than actively elaborate on the problem. Another important revelation is that when we expect to be able to find information again later on, we do not remember it as well as when we think it might become unavailable. Since search engines are continually available to us, we often feel we do not need to encode the information internally. Consequently, it changes our original memory of facts to a memory of ways to find the facts – we are learning what the computer knows and where to go to find information. Undoubtedly, future search tools should consider all of these effects.

# 3   Landmarks in Similarity Search Technology

## 3.1   Vector-Space Model

The elaboration of similarity search problems started in the *information retrieval* community, which has produced numerous concepts and technologies nowadays used in practical search engines, mostly processing text documents. Excellent textbook [8] provides a thorough and updated introduction to the key Infor-mation Retrieval principles behind search engines. It carefully surveys problems from parsing to indexing, from clustering to classification, from retrieval to rank-ing, and from user feedback to retrieval evaluation. All important concepts are carefully introduced and exemplified, slides for teaching are also available at the home-page[2] of the book.

However, the core of its success is the *vector-space model* with the *cosine similarity* to assess closeness of documents containing words. This certainly is a mature technology, based on efficient implementation through inverted files, and Google, Yahoo, and Microsoft (as well as several others) have proved its validity by enormous commercial success. This is also an excellent validation of the importance and usefulness of similarity in searching, though it only solves a specific, undoubtedly very important, form of similarity.

## 3.2   Metric Model

Probably the main stream of research towards a more generic and *extensible* form of similarity searching has, in the last 20 years, been developing around the concept of mathematical *metric space* [9]. Though the origins of the topic are older, the boom started in the 1990s with a sharp increase in number of publications and mainly citations to key scientific and technological achievements.

The metric space paradigm extends the range of possible similarity measures but at the same time loses the possible advantage of coordinate systems to define partitioning of search space. Since the similarity is in fact measured as dissimilarity, specifically a distance, the applied techniques are often designated as *distance searching*.

Several key publications summarize achievements in this area. The first survey [10] includes results till the year 2000. It presents known approaches in original taxonomy with the objective to discover core properties that would allow combination of existing principles to form future better proposals. An important contribution is a quantitative definition of the *intrinsic dimensionality* of metric data sets, which is strictly related to search complexity.

The second survey [11] divides existing methods for handling similarity search into two classes. The first class directly indexes objects based on distances (distance-based indexing), while the second is based on mapping to a vector space (mapping-based approach). However, the main part of this article is dedicated to a survey of distance-based indexing methods, and the mapping-based methods are only outlined. An important contribution is a presentation of a general framework for the execution of search operations, based on distances. The suggested algorithms for common types of queries – similarity range and nearest neighbors queries – operate on a generalized search hierarchy. Such algorithm finds a lot of applications in techniques used on hierarchical partitioning of data.

In 2006, a book named Similarity Search: The Metric Space Approach [12] presented the state-of-the-art in developing index structures and supporting technologies for searching complex data modeled as instances of a metric space. In Part I, the problem of metric searching is introduced and its importance with respect to other approaches is justified. It also presents examples of distance measures used for searching in diverse data collections. After defining similarity

---

[2] http://www.mir2ed.org/

queries, the basic partitioning principles are discussed in light of query execution. The rest of the introductory material is devoted to performance related issues, specifically techniques aimed at reducing the number of distance computations, metric space transformations, and concepts of approximate similarity search. Finally, a collection of analytic tools and methodologies specifically developed for metric index structures is discussed. The survey of existing search techniques in Part II is divided into three groups. In the first group, numerous proposals, prevalently suitable for searching in main memory, are systematically discussed, classified on applied partitioning paradigms. The second group of index structures assumes data to be stored on disc-like memories, so the input and output access costs are also considered. The third group of techniques surveys indexes able to trade some search precision, i.e. effectiveness, for substantial improvements in performance, i.e. efficiency. Finally, the parallel and distribute indexes, capable of inter, as well as the intra, query parallel execution are discussed.

At the home-page[3] of the book, you can also find a three-hour tutorial as well as a large collection of slides, covering the whole topic of the book – convenient as a teaching material. The metric searching problems are also considered in the last edition of the encyclopedic book by Hanan Samet [13] called Foundations of Multidimensional and Metric Data Structures.

### 3.3   Non-metric Models

It is obvious that the constraints of distance symmetry and triangle inequality of metric spaces are far too restrictive for many distance measures needed to model user perceived similarity in practice. An extensive recent survey [14] describes domains employing nonmetric functions for effective similarity search and methods for efficient non-metric similarity search. It is in agreement with psychological models of similarity, but the resulting measures are complex and fail to posses suitable geometric properties needed for partitioning and searching. Though the review gives many pointers to the state-of-the-art techniques for efficient (fast) non-metric similarity search, concerning both exact and approximate search, the issue is still a big open problem especially from the performance point of view. For this reason, the promising applications seem to be those not requiring online query response time.

### 3.4   Extending the History to Future

As pertinent to all complex systems, the future search must be born on the divergence of the *scale* and *determinism*. In particular, a more and more desirable property of any search system will be its ability to either handle growing amounts of work in a graceful manner, or to be readily enlarged. That means the scalability, in general, is going to be a more and more important issue. At the same time, the necessity of search processes being determined by an unbroken

---

[3] http://www.nmis.isti.cnr.it/amato/similarity-search-book/

chain of pre-defined steps – that is the determinism – is going to be less and less important in information seeking, sometimes even counterproductive. The effects of divergence of scale and determinism on the background of search structures development are illustrated in Fig. 1. The development starts with the well established centralized and parallel organizations, continues through the cutting-edge distributed and peer-to-peer approaches, and aims towards self-organizing search architectures.

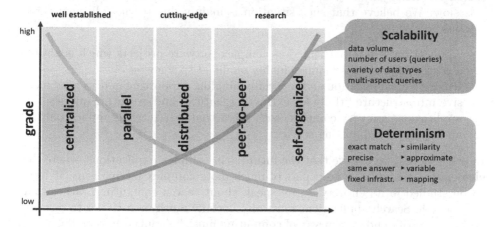

**Fig. 1.** Development trends in search structures

For example, current trends in a growing range of scale dimensions can be characterized by:

- exponential growth in data volume,
- number of users (queries) increasing fast,
- variety of data types, and emergence of various digital databases and libraries,
- multi-aspect (lingual, feature, modal) queries,

whereas the forms of softening effects of determinism can be exemplified by shifts from:

- exact match to similarity reasoning,
- precise query evaluation to approximate evaluation,
- unvaried answer to satisfactory answer,
- fixed search to personalized, context aware, and affective-based search,
- dedicated hardware to dynamic hardware mapping.

## 4   Towards a Better Similarity Search Technology

### 4.1   Why So Few Applications

Though a lot of progress has been done, the fact still is that the only successful application of similarity searching is the text similarity search through the vector-space model. Surprisingly, the attractive extensibility property – one system used for many applications – of the metric space approach to similarity searching, has not yet been fully exploited. There are examples of applications in image search, audio (music) processing, and several others, but significance, measured by commercial success, is marginal. Obviously, the technology is still developing and no doubt, better theories, paradigms, and technological proposals will appear in future. But the speed of spreading the more general similarity search technology by new applications – even with promising business models – is slow. We believe that such situation is mainly due to the following three reasons:

- Applications are implemented as complex software projects which is always a costly process as it requires highly qualified specialists;
- Running an application and especially building supporting indices need massive infrastructure fitfully for extracting features and executing queries;
- applications are very complex, with the actual search being applied only as an important supporting service.

In other words, we believe that the future is in complex applications, where the elementary search becomes a part of much more complex mechanisms. It need to be studied in its entirety, so we speak about search computing, rather that only search. Search applications must also become much more easier to develop and run. To this end, a new way of computing must be applied to searching, and the project-like approach substituted by the cloud computing paradigm. In the following, we elaborate more on these issues.

## 4.2   Search Computing

Searching is perhaps one of the oldest research topics of computer science. At the same time, search for information is among the most important applications of today's computing systems – second probably only to e-mails. However, in the future, searching will more and more frequently be integrated in complex applications and will slowly be disappearing as a stand-alone application.

The recent and prestigious Search Computing[4] project (SeCo) aims at building concepts, algorithms, tools, and technologies to support complex Web queries. It proposes a new paradigm for solving complex queries based on combining data extractions from distinct sources and data integration by means of specialized integration engines. Data extraction retrieves data from different sources, ordered based on local rankings, and data integration merges such inputs into result combinations, with an associated global ranking, such that combinations with the highest ranking are produced as fast as possible – a result combination represents the solution of a complex search problem. Thus, the search computing project has the ambitious goal of lowering the technological barrier required for building complex search applications, thereby enabling the development of many new applications which will cover relevant search needs.

---

[4] http://www.search-computing.it/

Business areas, research, and socio-economic challenges of search computing are discussed in the EC Multimedia Search Cluster White Paper[5]. For efficient and human-centric search, they suggest the following topics to be more closely explored:

– Multi-modal search enabling queries and interaction independently of the form in which the content is available;
– affective-based search taking into account both the users emotional state and the sentiment contained in multimedia documents;
– event-based representation and analysis as an efficient and user-centric approach for the annotation and retrieval of content;
– user experience aspects, including new generations of search interfaces (e.g., visual search interfaces, augmented reality, 3D browsing in virtual places) to enable novel forms of search input;
– content-aware network nodes extending search in the network with challenges related to discovery in the network;
– real-time approaches and architectures enabling to push information to users as fast as it is available, and at the same time maintaining the balance between quality, authority, relevance and timeliness of the content;
– content diversity in knowledge, providing the ability to identify and exploit the aspects that differentiate a piece of information from another;
– aggregation, mining, and data analysis enabling new forms of data processing for social network resources through interlinking data chunks under the Linked Open Data principles.

In addition to the research challenges, new business models are needed in applications where user-generated content plays an important role in order to allow both commercial exploitation and protection of user rights. In the classical *Intangible Benefits* models, free services are provided to users in exchange for their attention, loyalty, and information – then it is indirectly used in exchange for money. Profit does not derive directly from the use of the searching functionality but from attracting more customers to use a paid service that incorporates this functionality as an additional feature. On the other hand the *Monetary Benefits* model comes up in the majority of relationships where a transaction or a subscription process takes place and customers are required to pay in exchange of services or goods. This model is usually implemented through fixed transaction fees, referral fees, etc. It is generally believed that these models are not sufficient to cover all aspects of Search Computing and new ideas are expected to appear.

The future of similarity search is in applications, but at the same time, it is the applications which shape the problem of similarity searching. Besides the traditional *web search*, the following three business areas are promising for searching:

**Enterprise Search** - given the diversity of repositories present in enterprises, search engines became a unified method for accessing all corporate information, irrespective of the original repository. Current evolution of search-based applications recognizes the benefit of this approach and develops it further;

---

[5] http://avmediasearch.eu/white_paper_Search_computing

**Mobile Search** - the primary objective of mobile search is to enable people find either generic web or location-based information and services by entering a word phrase, preferably by voice, or an image from their phone;

**Social Search** - on top of the massive digital outcome produced by social media like Flickr, Facebook, or YouTube, social search takes a radically new shape. The traditional search areas are now extended by new modalities, such as time-stamps and geo-locations, but even tag co-occurrence or friendship links. The fact that users annotate and comment on the content in form of tags, ratings, preferences, etc. on daily bases, gives this data source an extremely dynamic nature that reflects events and the evolution of community focus.

### 4.3    Generic Model

Implementation of search applications is very sensitive to details and a weakness of one aspect results in failure of the whole system. In short, the system must not only be relevant, but mainly fast and simple to use. But these properties do not came easy, and a holistic approach is needed – the whole is more than a sum of its parts. To better understand the problem and the relationships between individual components of similarity search computing, consider the scheme in Fig. 2.

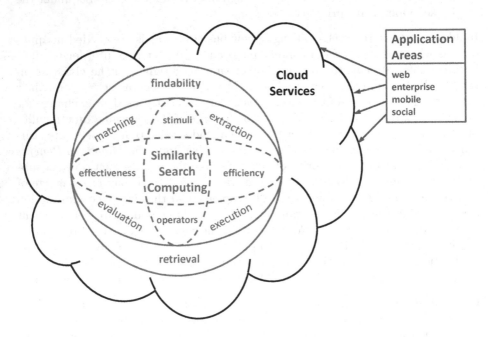

**Fig. 2.** Similarity search computing services

In principle, there are two primary axes. The horizontal one represents the *conceptual* view with effectiveness and efficiency properties on its extremes. The vertical axis constitutes the *implementation* view with stimuli and operations as extremes. In the following, we discuss these properties in more details.

*Effectiveness* represents abstract (possibly psychological) models of similarity; it concerns methods, methodologies, and models of similarity perception as well as techniques to assess their quality and user (application) satisfaction. On the other hand, *efficiency* represents generic infrastructures for high-performance similarity management. Specifically, it concerns architectures, models, computing infrastructures, and other software and hardware mechanisms able to support fast execution of computational intensive tasks. For example, efficiency includes parallelism, scalable and distributed architectures, self-organizing principles, fault tolerance, but also multi-processor systems, GPU's, etc.

*Stimuli* are representations of raw digital data and retain specific content. They are purpose-built representations of raw (complex digital) data objects encapsulating specific knowledge about the content of these objects. The representation of the knowledge should be maximally precise, but at the same time economic in space for storage purposes. *Operations* represent basic similarity search primitives, which typically come in several forms, such as the range search, nearest neighbor search, reverse nearest neighbor search, similarity joins, etc. But many other operations can be defined as well, for example a similarity ranking, classification, or merging. The objective is to develop similarity algebra of operations over digital data.

Developing a technology for similarity searching concerns generally four broad areas of extraction, matching, evaluation, and execution. They are conveniently located in Fig. 2 between extremes of the principle axes.

*Extraction* mainly concerns harvesting of stimuli, that is the computational methods of stimuli extraction, respecting specific types of raw data and stimuli. Depending on the volume of data, such process can be enormously time-consuming, therefore a special care must be taken from the efficiency point of view. For example, provided a feature extraction from an image takes half a second, it would take one year to finish this job for 60 million images – such number of images (or even more) is uploaded on Facebook just in one day.

*Matching* is a semantic specification of similarity measures and their assessment. It concerns the way how, given specific stimuli, required effectiveness is quantified. Obviously, given specific sets of stimuli, there are several ways of comparing them. For example, collections of numbers can be compared by the Euclidean distance on numeric vectors, the edit distance can be applied on numbers forming strings, and the Jaccard's Coefficient on sets.

*Evaluation* process aims at implementation of similarity models by means of available operations. The problem starts from users, their needs, unique attitude or context and requirements expressed through a specific model of similarity. It concerns a requirement specification tool (a language) to express needs as well as a communication environment, or interface, able to report results.

*Execution* should guarantee a performance-oriented evaluation of similarity operation transactions. It concerns a platform-dependent implementation techniques to achieve required performance. Recent studies confirm that users more and more emphasize the speed of search execution – recall falls dramatically as the searched collection increases in size, but the speed is important.

Considering the lower part of Fig. 2 again – that is the effectiveness, evaluation, operators, execution, and efficiency aspects – we have a search technology for *retrieval*. If we consider the upper part of the figure – that is the effectiveness, matching, stimuli, extraction, and efficiency – we have a technology which aims at *findability*, that is the ease with which information contained in complex data can be found.

In the past, most effort spent while building the similarity search technology concentrated on retrieval and the problems of findability were rather neglected. We consider this situation as one of the main reasons for the current similarity search problems. For this reason, we especially concentrate in the next section on the findability problems, that is the degree or quality to which a particular object is easy to discover by searching. We discuss these problems in context of *cloud computing* [15], which is a quite new communication paradigm especially promising for similarity searching. But it is also posing new research challenges, mainly for security of *outsourced* data in computer clouds.

## 5 Findability and Security in Similarity Cloud Services

A cloud is a platform or infrastructure that enables execution of code, services, or applications in a *managed* and *elastic* fashion. Cloud systems are not to be misunderstood as just another form of resource provisioning infrastructure. Multiple opportunities arise from principles of cloud infrastructures that will enable further types of applications and reduce development and provisioning time of different similarity search services. In general, cloud computing has particular characteristics that distinguish it from classical resource and service enabling environments, specifically:

**Scalability** - to support huge databases and very high request rates. Cloud systems are designed to scale-out, so that large scale is achieved by using large number of commodity servers, running in parallel. An effective scale-out system must balance load across servers and avoid bottlenecks;

**Elasticity** - to allow adding more capacity to a running system by deploying additional instances of each component, and shifting load to them. The capacity can also be reduced by analogy;

**Availability** - to provide high levels of useability and fault tolerance. In particular, cloud systems are often multi-tenant systems, which mean that a possible outage may affect many different applications, so the system must be very robust;

**Privacy** - outsourcing based on cloud computing is attractive as it offers *pay-as-you-go* low storage and processing costs as well as easy data access. However, care needs to be taken to safeguard data that is valuable or sensitive against unauthorized access.

Cloud computing poses a variety of challenges to conventional advanced similarity search technology. It is mostly due to the fact of the unprecedented scale and heterogeneity of forms of similarity and dramatically changing demands for infrastructure. It is a well known fact that multimedia objects, even such basic ones as images, can be accessible through many different content descriptors – if no descriptors are available, the complex Binary Large Object (BLOB) with no stimuli interpretations is not accessible (findable) and actually lost in the database.

Besides possible text or key-word annotations, numerous global and local descriptors can be extracted, creating a variety of domains for content-based retrieval through similarity. However, the feature extraction process is time consuming and requires not-trivial computational resources. On the other hand, to maximize the retrieval possibilities, to increase the object findability, many different descriptors should be extracted – the more descriptors extracted, the more infrastructure you need. Obviously the problem is even more complex (more demanding) when streams of objects, for example a sequence of images (video), are considered.

Cloud resources are potentially shared between multiple tenants. On the other hand, data might be sensitive, for example medical data, and cloud providers as well as users cannot always be trusted. Therefore, privacy and security becomes an important matter of concern with data outsourcing. All these aspects ask for rethinking of even current advanced search technology solutions.

In the following, we shortly discuss two selected problems associated with developing similarity searching services for clouds. In particular, we concentrate on execution platforms and architectures for multi-modal object findability and security issues in similarity search outsourcing.

### 5.1    Execution Platforms and Architectures for Multi-modal Object Findability

Findability of objects and the process of their retrieval are closely related. They are both dependent on defined object representation features, which make them distinct from and related to others according to a certain, possibly combined, criteria. In a physical environment, it is the size, shape, color, and location which set objects apart. In relational databases, we rely on attributes, while words as labels, links, or simply keywords have dominated the web search of text documents.

Findability is not a Search Engine Optimization (SEO) – SEO is only an aspect of findability and makes a BLOB findable for search engines, provided its features are extracted. Multimedia content should take advantage of descriptive meta-data, because it is the way objects can be found. Automated content tagging, annotation, classification, or simply content extraction is a difficult and time consuming process, because there are many (possibly complex) features, huge data repositories, and even continuous streams of data, that all require online processing. In this respect, MapReduce [16] like approaches should be considered and explored for possible application.

In general, such situation asks for new architectures of feature extraction for searching, dynamically exploiting powerful computational cloud platforms to support high availability of complex information needs over multimedia data.

## 5.2   Security Issues in Similarity Search Outsourcing

Strictly related to the issues of findability and elasticity is the concern of data protection and other potential security leaks arising from the fact that the resources are shared between multiple tenants and the location of the resources is potentially unknown. In particular, sensitive data or protected applications are critical for outsourcing approaches and need special treatment.

While essential security aspects are addressed by most existing multimedia processing tools, additional issues arise due to the specifics of cloud systems. They are in particular related to the complex management of the raw and extracted data, as well as the replication and distribution of the data in potentially worldwide resource infrastructures. The data should be protected in a form that addresses legislative issues with respect to data location, but at the same time, it should still be manageable by the system with minimum data transfers, which is very important for efficiency reasons.

In addition, the many usages of cloud systems and the variety of cloud types imply different security models and requirements of the user. As such, classical authentication models may be insufficient to distinguish between the aggregators/vendors and actual users. In similarity searching services, the data is to be revealed only to trusted users – not even to the service providers. Outsourcing should offer scalability and low initial investment for the data owner, while still providing the essential service, that is fast retrieving most similar data objects to a query reference.

In any case, the privacy of sensitive or otherwise confidential data must be guaranteed. A possible solution might be to find transformations that would change the data prior to submitting to a service provider. Search might then be performed on the changed data, having actually a wrong or even no meaning without knowing the (owner-defined) transformation key.

In summary, new security governance models and processes are required that cater for specific issues of similarity computing arising from the cloud model. A pioneering work in this direction is [17].

## 6   Conclusions

Similarity search in collections of complex objects has proved useful, but stays prevalently on the level of the text search. Though several other applications are also already operational [18] through projects like MUFIN [4], there are still problems which prevent faster development.

First, the similarity search is still expensive considering both the processing time and the storage costs, thus better paradigms and techniques for feature extraction and query execution, properly exploiting up-to-date computing infrastructures, are still needed. Second, existing software is typically not available in a form of simple and ready-to-use packages and developing of applications requires project-like approaches. These are costly, consuming extensive human and infrastructure resources, which are scarce and not always available.

A promising alternative to be exploited in the future is to develop similarity search services based on the cloud computing paradigm. The expected positive properties can mainly be attributed to the specific characteristics of cloud computing systems. In theory, these are practically infinitely scalable. They provide infrastructures for platforms, a platform for applications, or directly applications themselves as services. Clouds can be seen as a generalized platform of a fully outsourced ICT service for an organization. Clouds also shift the costs for a business and avoid costly asset acquisitions.

However, such approach poses new research challenges, some of which have been outlined in this paper.

**Acknowledgments.** This research was partially supported by the Czech Science Foundation project number P103/12/G084. The participation at the SISAP 2012 Conference was supported by the FIELDS INSTITUTE Research in Mathematical Science, Toronto, Canada.

# References

1. Larkey, L., Markman, A.: Processes of similarity judgment. Cognitive Science 29, 1061–1076 (2005)
2. Vosniadou, S., Ortony, A.: Similarity and Analogical Reasoning. Cambridge University Press (2003)
3. Kurdek, L., Schnopp-Wyatt, D.: Predicting relationship commitment and relationship stability from both partners' relationship values: Evidence from heterosexual dating couples. Personality and Social Psychology Bulletin 23(10), 1111–1119 (1997)
4. Zezula, P.: Multi Feature Indexing Network MUFIN for Similarity Search Applications. In: Bieliková, M., Friedrich, G., Gottlob, G., Katzenbeisser, S., Turán, G. (eds.) SOFSEM 2012. LNCS, vol. 7147, pp. 77–87. Springer, Heidelberg (2012)
5. Morville, P.: Ambient findability - what we find changes who we become. O'Reilly (2005)
6. Morville, P., Callender, J.: Search Patterns - Design for Discovery. O'Reilly (2010)
7. Sparrow, B., Liu, J., Wegner, D.M.: Google effects on memory: Cognitive consequences of having information at our fingertips. Science 333, 776–778 (2011)
8. Baeza-Yates, R., Ribeiro-Neto, B.: Modern Information Retrieval - the concepts and technology behind search, 2nd edn. ACM Press Books (2011)
9. O'Searcoid, M.: Metric Spaces. Springer (2006)
10. Chávez, E., Navarro, G., Baeza-Yates, R., Marroquín, J.: Searching in metric spaces. ACM Comput. Surv. 33(3), 273–321 (2001)
11. Hjaltason, G., Samet, H.: Index-driven similarity search in metric spaces. ACM Trans. Database Syst. 28(4), 517–580 (2003)

12. Zezula, P., Amato, G., Dohnal, V., Batko, M.: Similarity Search: The Metric Space Approach. Advances in Database Systems, vol. 32. Springer (2006)
13. Samet, H.: Foundations of Multidimensional And Metric Data Structures. Series in Data Management Systems. Morgan Kaufmann (2006)
14. Skopal, T., Bustos, B.: On nonmetric similarity search problems in complex domains. ACM Comput. Surv. 43(4), 34 (2011)
15. Sosinsky, B.: Cloud Computing Bible. Wiley (2011)
16. Dean, J., Ghemawat, S.: Mapreduce: simplified data processing on large clusters. Comm. ACM 51(1), 107–113 (2008)
17. Yiu, M., Assent, I., Jensen, C., Kalnis, P.: Outsourced similarity search on metric data assets. IEEE Trans. Knowl. Data Eng. 24(2), 338–352 (2012)
18. Novak, D., Batko, M., Zezula, P.: Generic similarity search engine demonstrated by an image retrieval application. In: Proceedings of the 32nd Annual International ACM SIGIR Conference on Research and Development in Information Retrieval, Boston, MA, USA, July 19-23, p. 840 (2009)

# Snake Table: A Dynamic Pivot Table for Streams of k-NN Searches

Juan Manuel Barrios[1], Benjamin Bustos[1], and Tomáš Skopal[2],*

[1] KDW+PRISMA, Department of Computer Science, University of Chile
{jbarrios,bebustos}@dcc.uchile.cl
[2] SIRET Research Group, Faculty of Mathematics and Physics,
Charles University in Prague
skopal@ksi.mff.cuni.cz

**Abstract.** We present the Snake Table, an index structure designed for supporting streams of $k$-NN searches within a content-based similarity search framework. The index is created and updated in the online phase while resolving the queries, thus it does not need a preprocessing step. This index is intended to be used when the stream of query objects fits a snake distribution, that is, when the distance between two consecutive query objects is small. In particular, this kind of distribution is present in content-based video retrieval systems, when the set of query objects are consecutive frames from a query video. We show that the Snake Table improves the efficiency of $k$-NN searches in these systems, avoiding the building of a static index in the offline phase.

**Keywords:** Similarity Search, Metric Indexing, Multimedia Information Retrieval, Content-Based Video Retrieval.

## 1 Introduction

In this paper we present the Snake Table, which is an indexing structure designed for supporting streams of $k$-NN searches. The index is intended to be used when the stream of query objects fits a "snake distribution", which we define formally in this work. This kind of distribution is usually present in content-based video retrieval systems, when the set of query objects corresponds to consecutive frames from a query video. In particular, we evaluate the Snake Table on a Content-based Video Copy Detection (CBVCD) system where the query objects present a snake distribution. Unlike most of the index structures, the Snake Table is a session-oriented and short-lived index.

An existing indexing structure with similar objectives and properties is called the D-cache [9]. In this work, we show that the D-cache suffers from high internal realtime complexity making it unviable to use in a CBVCD system or other systems with computationally inexpensive (i.e., fast) distance functions (like

---

* This research has been supported in part by Czech Science Foundation project GA CR 202/11/0968.

G. Navarro and V. Pestov (Eds.): SISAP 2012, LNCS 7404, pp. 25–39, 2012.

Manhattan distance or Euclidean distance). Also, we compare the Snake Table with D-cache and LAESA index, and we show that Snake Table achieves the best performance.

The structure of the paper is as follows. Section 2 gives a background of metric spaces and efficiency issues. Section 3 reviews the related work. Section 4 gives the definition of the Snake Table and snake distribution. Section 5 presents the experimental results. Finally, Section 6 concludes the paper and outlines some future work.

## 2   Background

Let $\mathcal{M} = (\mathcal{D}, d)$ be a metric space [11]. Given a collection $\mathcal{R} \subseteq \mathcal{D}$, and a query object $q \in \mathcal{D}$, a range search returns all the objects in $\mathcal{R}$ that are closer than a distance threshold $\epsilon$ to $q$, and a nearest neighbor search ($k$-NN) returns the $k$ closest objects to $q$ in $\mathcal{R}$.

For improving efficiency in metric spaces, Metric Access Methods (MAMs) [3] are index structures designed to efficiently perform similarity search queries. MAMs avoid a linear scan over the whole database by using the metric properties to save distance evaluations. Given the metric space $\mathcal{M}$, the object-pivot distance constraint [11] guarantees that:

$$\forall a, b, p \in \mathcal{D}, \ |d(a,p) - d(p,b)| \leq d(a,b) \leq d(a,p) + d(p,b) \qquad (1)$$

One index structure that uses pivots for indexing is the Approximating and Eliminating Search Algorithm (AESA) [10]. It first computes a matrix of distances between every pair of objects $x, y \in \mathcal{R}$. The structure is simply an $|\mathcal{R}| \times |\mathcal{R}|$ matrix holding the distances between every pair (due to the symmetry property of $d$ only a half of the matrix needs to be stored). The main drawback of the AESA approach is the quadratic space of the matrix. Linear AESA (LAESA) [8] gets around this problem by selecting a set of pivots $\mathcal{P} \subseteq \mathcal{R}$. The distance between each pivot to every object is calculated and stored in a $|\mathcal{R}| \times |\mathcal{P}|$ distance matrix, also known as the *pivot table*. LAESA reduces the required space compared to AESA, however an algorithm for selecting a good set of pivots is required [2].

Given a query object $q$ (not necessarily in $\mathcal{R}$), the similarity search algorithm first evaluates the distance $d(q, p)$ for each pivot $p \in \mathcal{P}$, then scans $\mathcal{R}$ and for each $r \in \mathcal{R}$ it evaluates the lower bound function $\text{LB}_{\mathcal{P}}$:

$$\text{LB}_{\mathcal{P}}(q, r) = \max_{p \in \mathcal{P}} \{|d(q, p) - d(r, p)|\} \qquad (2)$$

Note that $\text{LB}_{\mathcal{P}}$ can be evaluated efficiently because $d(q, p)$ is already calculated and $d(r, p)$ resides in the pivot table. In the case of range searches, if $\text{LB}_{\mathcal{P}}(q, r) > \epsilon$ then $r$ can be safely discarded because $r$ cannot be part of the search result. In the case of $k$-NN searches, if $\text{LB}_{\mathcal{P}}(q, r) \geq d(q, o^k)$ then $r$ can be safely discarded, where $o^k$ is the current $k^{\text{th}}$ nearest neighbor candidate to $q$. If $r$ could not be discarded, then the actual distance $d(q, r)$ must be evaluated to decide whether or not $r$ is part of the search result.

The efficiency of some MAM in $\mathcal{M}$ is related to: 1) the number of distance evaluations that are discarded when it performs a similarity search; and 2) the internal cost for deciding whether some distance can be discarded or not. A similarity search using any MAM will be faster than a linear scan when the time saved due to the discarded distances is greater than the time spent due to the internal cost. For example, in the case of LAESA, the internal cost for a similarity search comprises the evaluation of $d(q, p)$ for each pivot $p$ in $\mathcal{P}$, and the evaluation of $\text{LB}_\mathcal{P}(q, r)$ for each object $r$ in $\mathcal{R}$, thus it increases linearly with $|\mathcal{P}|$. The amount of distances discarded by LAESA depends on the size and quality of $\mathcal{P}$ and on the metric space itself.

In order to analyze the efficiency that any MAM can achieve in a collection $\mathcal{R} \subseteq \mathcal{D}$ for some metric space $\mathcal{M} = (\mathcal{R}, d)$, Chávez et al. [3] propose to analyze the histogram of distances of $d$. A histogram of distances is constructed by evaluating $d(a, b)$ for a random sample of objects $a, b \in \mathcal{R}$. The histogram of distances reveals information about the distribution of objects in $\mathcal{M}$. Given a histogram of distances for $\mathcal{M}$, the intrinsic dimensionality $\rho$ is defined as $\rho(\mathcal{M}) = \frac{\mu^2}{2\sigma^2}$, where $\mu$ and $\sigma^2$ are the mean and the variance of histogram of distances for $\mathcal{M}$. The intrinsic dimensionality estimates the efficiency that any MAM can achieve in $\mathcal{M}$, therefore it tries to quantify the difficulty in indexing a metric space. A histogram of distances with small variance (i.e., a high value of $\rho$) means that the distance between any two objects $d(a, b)$ with high probability will be near $\mu$, thus the difference between any two distances with high probability will be a small value. In that case, for most of the pivots the lower bound from Eq. 1 will probably become useless at discarding objects. Increasing the number of pivots will improve the value of the lower bounds, however the internal cost of the MAM will also increase.

## 3   Streams of k-NN Searches

MAMs can be classified as static or dynamic depending on how they manage the insertion or deletion of objects in $\mathcal{R}$ during the online phase. A dynamic MAM can update its structures to add or remove any object, hence it can remain online even for growing collections. Usually, the tree-based MAMs, like the M-Tree, are dynamic. A static MAM cannot manage large updates in its structures, thus after many modifications of $\mathcal{R}$ the whole indexing structure must be rebuilt. LAESA can manage the insertion or deletion of objects and pivots [7] by adding or removing rows or columns from the pivot table. However, depending on its actual implementation, LAESA is usually a static index, mainly because the pivot table might not support to dynamically modify its structure. In that case, a new table is required, copying values from the old table to the new one, evaluating the distances for new objects or pivots, and discarding the old table. Also, after many modifications in $\mathcal{R}$ the set of pivots can begin to perform poorly and a new set of pivots should be selected.

Most of the MAMs are designed to be created during the offline phase, that is, a time-expensive process creates the index structure prior to resolve any search.

It is expected that the MAM will resolve many similarity searches, amortizing its creation time, but no information is a priori known about the query objects that will be resolved afterwards. In the online phase, the MAM efficiently receives and resolves any similarity search that may proceed from different sources and users. All the searches share the same MAM, and the MAM should achieve good performance for any search.

However, depending on the domain, the query objects may have some special properties that can be exploited to improve the performance of the MAM. In particular, frame-based CBVCD systems usually divide a query video into shots and many keyframes are extracted. A similarity search is then performed for consecutive keyframes, thus it can be expected that two consecutive keyframes will frequently be similar. In the case of interactive Content-based Multimedia Information Retrieval (CBMIR) systems, a user starts a search with some example or some tags, a $k$-NN search is performed and the answers are shown, then iteratively the user selects a new query object among those shown, and a new search is performed refining the results. Because the new queries are selected from the answers of a previous search, it can be expected that two consecutive query objects will be similar.

### 3.1  Related Work – D-Cache and D-File

To take advantage of the online indexing process and a stream of correlated queries, there is a recently proposed structure called D-file [9]. The D-file is the database file itself accompanied by a main-memory structure, called the D-cache. The D-cache stores the evaluated distances $d(q_i, o_j)$ while processing queries in the stream. When the $n^{\text{th}}$ query in the stream is processed, the D-cache calculates a lower-bound distance for $d(q_n, o_j)$ evaluating the distance from $q_n$ to previous $q_i$ and treating the previous queries as pivots. Hence, if the calculated lower bound is large enough, $o_j$ can be discarded without evaluating the actual distance to $q_n$. D-cache content is modeled as a sparse dynamic pivot table, where each table row is constructed with the stored distances. If there are not enough distances stored in the D-cache, some rows are incomplete, resulting in zeros on some cells. Using the reconstructed rows, the D-cache tries to filter out each database object using the same approach as a regular pivot table. The D-file does not need an offline indexing step, as the D-cache is being built during query processing. As the D-cache uses the previously processed queries as dynamic pivots, the authors recommend that previous queries should be as close to the current query as possible.

The D-cache is implemented with: 1) a fixed-size hash table that stores triplets $(q_i, o_j, d(q_i, o_j))$; 2) a hash function $h(q_i, o_j)$ for accessing the bucket where a triplet is stored; 3) a collision interval, for searching a near available bucket when some triplet is mapped into an already used bucket; and 4) a replacement policy, that decides whether or not a new triplet should replace an old triplet when a collision occurs and there is not an available bucket in the collision interval.

In CBVCD systems, a similarity search is performed for consecutive keyframes, thus it can be expected that the D-cache will achieve high performance. However, as we show in the experimental section, the D-cache suffers from high internal realtime complexity rendering it unviable to use in a CBVCD system. The main problem arises when the distance function is not time-expensive. In that case, the internal complexity associated with the hash function and collision resolution dominates the search times. In order to solve this problem, we introduce the Snake Table that preserves the idea and advantages of D-file and D-cache, but exhibits lower internal complexity.

## 4   Snake Table

In this work we propose a new dynamic indexing structure, called a Snake Table, which is designed to: 1) improve the search time for streams of queries where consecutive query objects are similar; and 2) keep its internal complexity low to be applied in systems that use fast distance functions, like CBVCD systems and interactive CBMIR that use global descriptors and Minkowski distances.

The life cycle of the Snake Table is as follows: First, when a new session is created, an empty Snake Table is created and associated with it. When a query object $q_1$ is received, a $k$-NN search is performed. The distances between $q_1$ and the objects in the collection are added to the Snake Table, and the result is returned. Then, when a new query object $q_i$ is received, a $k$-NN is performed using the previous query objects $q_1, ..., q_{i-1}$ as pivots to accelerate the search. Finally, when the session ends, the Snake Table is discarded. Therefore, like D-cache and unlike most of MAMs, the Snake Table is a session-oriented and short-lived MAM.

The Snake Table is implemented with a fixed-size $|\mathcal{R}| \times p$ matrix used as a dynamic pivot table. As in LAESA, the $j^{\text{th}}$ row in the dynamic pivot table represents the object $o_j$ in $\mathcal{R}$ and contains the distances between $o_j$ and up to $p$ previously processed query objects. However, each cell in the $j^{\text{th}}$ row of the table contains a pair $(q, d(q, o_j))$ for some query object $q$ (not necessarily in order). When processing a new query object $q_i$, the lower bound $\text{LB}_{\mathcal{P}}(q_i, o_j)$ for the distance $d(q_i, o_j)$ is calculated (see Eq. 2), with $\mathcal{P}$ dynamically determined by the query objects and distances in the $j^{\text{th}}$ row. The object $o_j$ is discarded when $\text{LB}_{\mathcal{P}}(q_i, o_j)$ is greater than the distance between $q_i$ and its current $k^{\text{th}}$ nearest neighbor candidate (obtained between $o_1$ and $o_{j-1}$). If $o_j$ is not discarded, the actual distance $d(q_i, o_j)$ is evaluated, added to some cell in the $j^{\text{th}}$ row, and the NN candidates are updated when $o_j$ is closer than the current $k^{\text{th}}$ NN to $q_i$.

We present three different replacement strategies to assign a distance $d(q_i, o_j)$ to one of the $p$ cells in the $j^{\text{th}}$ row:

1. Each query $q_i$ picks a column in round-robin mode, i.e., the distance $d(q_i, o_j)$ is stored in the $(i \bmod p)$ column of $j^{\text{th}}$ row, eventually replacing the stored distance $d(q_{i-p}, o_j)$. If the distance was not evaluated because it was discarded by $\text{LB}_{\mathcal{P}}(q_i, o_j)$ there are two options: 1) its corresponding cell is

either updated with an $\infty$ distance; 2) the cell is left unmodified but before any read the query stored in the cell is matched with the last query for that column (the experimental section uses the latter). With this strategy, each row is sparse, containing at most $p$ distances between $d(q_{i-p}, o_j)$ and $d(q_i, o_j)$.

2. The distance $d(q_i, o_j)$ is compared to every distance in $j^{\text{th}}$ row and the highest distance in the row is replaced. With this strategy, each row stores $p$ unsorted distances between $d(q_1, o_j)$ and $d(q_i, o_j)$.

3. Each distance $d(q_i, o_j)$ is stored in a cell chosen in an independent round-robin for every row. With this strategy, every row compactly stores the last $p$ evaluated distances for $o_j$ replacing the old ones. $\text{LB}_{\mathcal{P}}$ starts its evaluation from the last stored distance and goes backwards, therefore favoring the most recent stored distances.

D-cache uses a combination of strategies 1 and 2. It always replaces an old distance (older than $q_{i-p}$), but if there is not an old distance in the collision interval, then it replaces the worst distance, defined as the distance closer to the median (or to some predefined percentile of distances). Note that a very high distance can achieve better discarding performance than a medium distance (as used in D-cache strategy), however they are unlikely to appear in a snake distribution. In order to reduce the internal complexity of strategy 2 this case is not considered.

For strategy 1, distances $d(q_i, q_j)$ with $j \in \{i - p, ..., i - 1\}$ are calculated and stored in memory at the beginning of every search. For strategies 2 and 3, distances $d(q_i, q_j)$ with $j \in \{1, ..., i - 1\}$ are calculated on-demand by $\text{LB}_{\mathcal{P}}$. Note that the internal complexity of strategy 3 is slightly higher than strategy 1, because it needs to manage an independent index for each row to mark the position of the last stored distance.

The performance achieved by these three replacement strategies are compared in the experimental section. However, despite the replacement strategy used by the Snake Table, the overall performance of the Snake Table mainly depends on the distribution of the query objects.

## 4.1  Snake Distribution

The Snake Table is intended to be used when the query objects in a stream fit a "snake distribution". Intuitively, we define that a set of objects fits a snake distribution when the distance between two consecutive objects in the stream is small compared to the average distance between any two objects (see Fig. 1). To measure and compare this fit, we define an indicator using the histogram of distances of $d$ for $\mathcal{Q}$ and $\mathcal{R}$.

Because the area of the histogram of distances is normalized to 1, the histogram can be seen as a probability distribution of the distances calculated by $d$. Then, we define the cumulative distribution in a similar way as in probabilities:

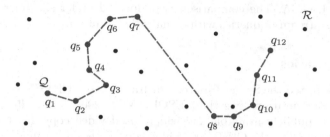

**Fig. 1.** Stream of queries $\mathcal{Q}=\{q_1, ..., q_{12}\}$ with a snake distribution: most of distances $d(q_i, q_{i+1})$ are smaller than $d(x, y)$ for randomly selected pairs $x,y$ in $\mathcal{R}$

**Definition 1.** *(Cumulative Distribution) Let $H$ be a normalized histogram of distances, the Cumulative Distribution of Distances $F : \mathbb{R}^+ \to [0,1]$ is defined as:*

$$F(x) = \int_0^x H(t)\, dt$$

For comparing the distribution of distances of two sets of objects, we compare their cumulative distributions:

**Definition 2.** *(Difference $\Delta$) Let $F_1$ and $F_2$ be two cumulative distributions, the difference $\Delta$ between $F_1$ and $F_2$ is defined as:*

$$\Delta(F_1, F_2) = \int_0^\infty F_1(t) - F_2(t)\, dt$$

The Difference $\Delta$ is meaningful only when both $F_1$ and $F_2$ originate from the same metric space. $\Delta(F_1, F_2)$ is greater than zero when the distances in $F_1$ are smaller than the distances in $F_2$.

**Definition 3.** *(Snake Distribution) Let $\mathcal{M} = (\mathcal{D}, d)$ be a metric space, let $\mathcal{R} \subset \mathcal{D}$ be the collection of objects, and let $\mathcal{Q} \subset \mathcal{D}$ be a set of $m$ query objects $\mathcal{Q} = \{q_1, ..., q_m\}$. Let $F$ be the cumulative distribution of $d(x, y)$ with random pairs $x, y \in \mathcal{Q} \cup \mathcal{R}$, $p$ be a number between 1 and $m-1$, and $F_\mathcal{Q}^p$ be the cumulative distribution of $d(q_i, q_{i-p}) \ \forall \ i \in \{p+1, ..., m\}$. $\mathcal{Q}$ fits a snake distribution of order $p$ if $\Delta(F_\mathcal{Q}^p, F) > s$, for some threshold value $s \in \mathbb{R}^+$.*

Note that when both $\mathcal{Q}$ and $\mathcal{R}$ are random samples of $\mathcal{D}$ without any special ordering (i.e., the $i^{\text{th}}$ sample does not depend on previous samples), then $\Delta(F_\mathcal{Q}^p, F) \approx 0$. When a distribution fits a Snake Distribution of order 1 to $p$ then a Snake Table can be created with a sliding window containing up to $p$ query objects.

## 5    Experimental Evaluation

In this section we evaluate the performance of the Snake Table with the three presented strategies, and we compare them with the performance achieved by

D-cache and LAESA. The comparison is performed under six different metric spaces, with a stream of queries with a snake distribution.

## 5.1  Preliminaries

**Dataset.** We tested the Snake Table on our frame-based CBVCD system [1] using different configurations over the MUSCLE-VCD-2007 dataset [5]. MUSCLE-VCD-2007 is a publicly available and widely-used video copy database, and it was the corpus used at the CIVR 2007 video copy detection evaluation. The reference collection is composed of 101 videos, 59 hours total length, and the query videos are divided into collections called ST1 and ST2. ST2 has three videos with a total length of 45 minutes. The ST2 collection contains 21 video excerpts copied from a video in the reference collection. Each copied excerpt may have some transformations like blur, flip, subtitles, zoom, insertion of logo, noise, etc. Every video in the dataset has 25 fps.

For the present evaluation, each reference video and each video in ST2 is partitioned into short fixed-length segments of 1 second. For each segment, four global descriptors are calculated: the Edge Histogram (EH), captures the spatial distribution of edges in a frame [6]. We used 10 orientations and 8-bits linear quantization, producing a vector of 160 bytes. The Ordinal Measurement (OM), captures the spatial distribution of intensities in a frame [4]. We used $9 \times 9$ blocks, producing a vector of 81 bytes. The Color Histogram (CH), divides a frame into 4 horizontal slices. Each slice calculates a histogram of 16 bins for R, G, and B channels, and each bin with 8-bits linear quantization, producing a vector of 192 bytes. The Keyframe (KF), reduces the frame to $11 \times 9$ pixels and uses the value for each pixel, producing a vector of 99 bytes. These descriptors are calculated for all the frames in a segment and then averaged.

$\mathcal{R}$ is the set of reference segments ($|\mathcal{R}|$=211,479 segments), and $\mathcal{Q}$ is the set of query segments ($|\mathcal{Q}|$=2,692 segments). The correct answer for a segment $q \in \mathcal{Q}$ is the reference segment $r_q \in \mathcal{R}$ for which $q$ is a copy. Because we stated that there is only one correct answer, the mean average precision (MAP) corresponds to the average of the inverse of the ranks of $r_q$, for all copied segment in $\mathcal{Q}$.

**Configurations.** We test six configurations, each one defining a distance function $d(r, s)$ between the video segments $r$ and $s$. The distances are based on linear combinations of $L_1$ (Manhattan) distance between descriptors, where $L_1(\boldsymbol{x}, \boldsymbol{y}) = \sum_{i=1}^{n} |x_i - y_i|$ for $n$-dimensional vectors $\boldsymbol{x}$ and $\boldsymbol{y}$:

1. **OM**: compares OM descriptors $d(r, s)$=$L_1(\mathrm{OM}(r), \mathrm{OM}(s))$.
2. **KF**: compares KF descriptors $d(r, s)$=$L_1(\mathrm{KF}(r), \mathrm{KF}(s))$.
3. **EH**: compares EH descriptors $d(r, s)$=$L_1(\mathrm{EH}(r), \mathrm{EH}(s))$.
4. **CH**: compares CH descriptors $d(r, s)$=$L_1(\mathrm{CH}(r), \mathrm{CH}(s))$.
5. **ECK**: weighted combination of EH, CH, and KF descriptors:

$$d(r, s) = 0.6 \times L_1(\mathrm{EH}(r), \mathrm{EH}(s)) \times \frac{1}{7996}$$
$$+ 0.2 \times L_1(\mathrm{CH}(r), \mathrm{CH}(s)) \times \frac{1}{6219}$$
$$+ 0.2 \times L_1(\mathrm{KF}(r), \mathrm{KF}(s)) \times \frac{1}{24721}$$

Each left factor (0.6, 0.2, 0.2) is the weight in the combination, and each normalization factor (7996, 6219, 24721) is the maximum distance value for the respective distance.

6. **EK3**: temporal combination of EH and KF descriptors:

$$g(r, s) = 0.5 \times L_1(\mathrm{EH}(r), \mathrm{EH}(s)) \times \frac{1}{7996}$$
$$+ 0.5 \times L_1(\mathrm{KF}(r), \mathrm{KF}(s)) \times \frac{1}{24721}$$
$$d(r_i, s_j) = \frac{1}{3} \left[ g(r_{i-1}, s_{j-1}) + g(r_i, s_j) + g(r_{i+1}, s_{j+1}) \right]$$

Where $r_{i-1}$ and $r_{i+1}$ are the previous and the next segments of $r_i$ in a video.

**Indexes.** We compare the efficiency of six indexes with $p$ pivots (either static pivots for LAESA or dynamic pivots for D-cache and the Snake Table), where $p$ varies between 1 and 20:

1. **D-cache**: It uses a hash table with fixed size $|\mathcal{R}| * p$, the collision interval to the minimum (1), and the hash function is $h(q_i, o_j) = (rnd_i * rnd_j) \bmod (|\mathcal{R}| * p)$, where $rnd_i$ and $rnd_j$ are unique random IDs assigned to each object. We checked that the hash function generates a uniform distribution through the whole table, producing almost no collisions.

2. LAESA: Following its definition, LAESA does not require any information of the query objects, but for a fair comparison, we allow LAESA to use $\mathcal{Q}$ in the selection process. **LaesaR** chooses $p$ static pivots from $\mathcal{R}$, and **LaesaQ** chooses $p$ static pivots from $\mathcal{Q}$. Both selections are performed using the SSS algorithm [2]. Four different sets are selected and the average value of $LB_{\mathcal{P}}$ is calculated for each one by sampling pairs from $\mathcal{Q} \times \mathcal{R}$. The set of pivots with higher average $LB_{\mathcal{P}}$ is finally selected while the other sets are discarded.

3. Snake Table: We test the three strategies depicted in Section 4. **SnakeV1** uses a sparse row with the last $p$ queries, **SnakeV2** uses an unsorted row discarding the highest distance, and **SnakeV3** uses a compact row with the last $p$ evaluated distances.

## 5.2 Experiments

Table 1 shows for the different configurations the total time spent by a linear scan (in seconds), the achieved MAP, and some indicators for the metric space. The

**Table 1.** Effectiveness and efficiency for the base configurations

|  | Time | MAP | max | $\mu$ | $\sigma$ | $\rho$ | $H_d$ |
|---|---|---|---|---|---|---|---|
| **Group 1** |  |  |  |  |  |  |  |
| OM | 282 s. | 0.125 | 3285 | 1489 | 416 | 6.4 |  |
| KF | 304 s. | 0.509 | 24721 | 7264 | 2636 | 3.8 |  |
| **Group 2** |  |  |  |  |  |  |  |
| EH | 541 s. | 0.639 | 7996 | 3198 | 751 | 9.1 |  |
| CH | 501 s. | 0.482 | 6219 | 3661 | 970 | 7.1 |  |
| **Group 3** |  |  |  |  |  |  |  |
| ECK | 1258 s. | 0.646 | 0.888 | 0.416 | 0.09 | 11.4 |  |
| EK3 | 2214 s. | 0.732 | 0.870 | 0.347 | 0.08 | 10.2 |  |

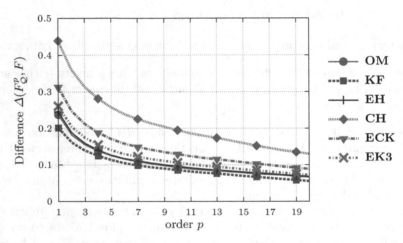

**Fig. 2.** Snake distribution of order $p \in \{1, ..., 20\}$ for the six configurations

histogram of distances was created by evaluating $d(x, y)$ with pairs $x$, $y$ sampled from $\mathcal{Q} \cup \mathcal{R}$. The configurations are split into three groups. Group 1 (**OM** and **KF**) contains the configurations where the linear scan takes less amount of time. Group 2 (**EH** and **CH**) contains the configurations where linear scan takes about twice as much time as Group 1. Group 3 (**ECK** and **EK3**) contains the configurations in which the linear scan is slower by one order of magnitude. In the following experiments, the performance of each index is presented as a ratio with the performance of the linear scan for that configuration. The MAP achieved by each configuration is also shown in the table. The configuration with best detection effectiveness is **EK3** followed by **ECK**. Also, **KF** shows a promising tradeoff between effectiveness and efficiency.

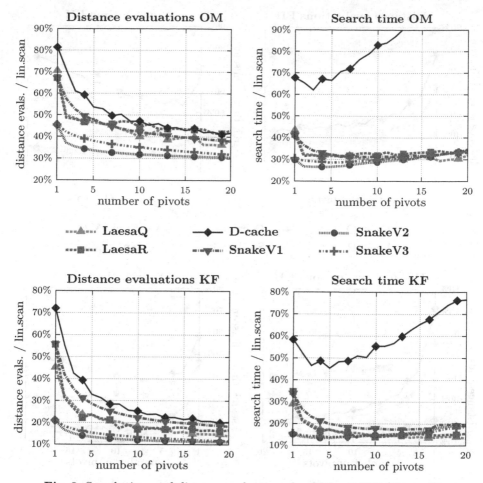

**Fig. 3.** Search time and distance evaluations for **OM** and **KF** (Group 1)

**Snake Distribution.** Figure 2 depicts the snake distribution of order $p$ for the six configurations. The value of difference $\Delta(F_{\mathcal{Q}}^p, F)$ for $p \in \{1, ..., 20\}$ is shown. The six configurations present a difference $\Delta$ higher than zero, hence the streams of queries have a snake distribution (distances between $q_i$ and $q_{i-p}$ are smaller than distances between random sampled pairs). The first orders show good a fit for the six configurations, but as $p$ increases, the snake distributions tend to disappear. As shown in the following experiments, the different configurations present satisfactory results for roughly between 1 and 5 pivots.

**Group 1.** Figure 3 shows the efficiency achieved by the six indexes for the **OM** and **KF** configurations varying the number of pivots from 1 to 20. It shows the amount of distances evaluations as a proportion of the evaluations required by the linear scan (i.e., a fraction of $|\mathcal{Q}|*|\mathcal{R}|$). This value includes the distances between query and pivots but does not include the distance required for pivot

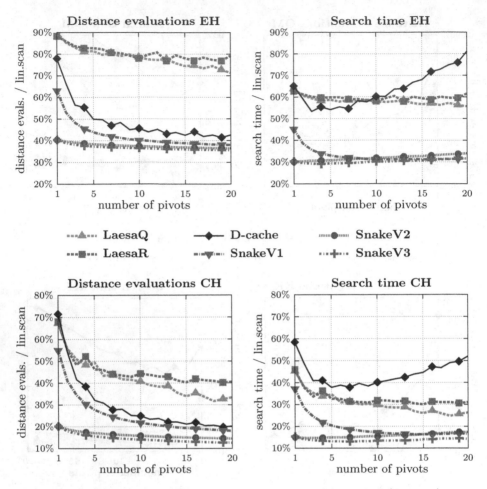

**Fig. 4.** Search time and distance evaluations for **EH** and **CH** (Group 2)

selection in LAESA. It also shows the search time as a proportion of the time spent by a linear scan. The instability in LAESA indexes for consecutive $p$ is due to the random-base pivot selection. To reduce this issue we calculate three different sets of pivots with SSS and the average value for evaluated distances and search time is presented.

The disparity between saved distances and saved time reveals that **D-cache** suffers from high internal complexity at these two configurations: while most of the distance computations are discarded, the search time increases even beyond the time required by a linear scan in **OM**. Both **LaesaR** and **LaesaQ** (i.e., static pivots) perform better than **D-cache**. On average, **LaesaQ** performs slightly better than **LaesaR** in both experiments. The Snake Table achieves the best performance by its combination of good pivot selection (due to the snake distribution) and its low internal complexity. Between the different strategies, the **SnakeV2** (i.e., replacement of the highest distance) achieves the best performance.

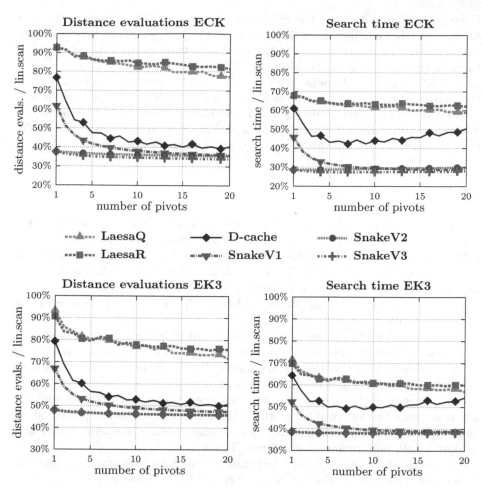

**Fig. 5.** Search time and distance evaluations for **ECK** and **EK3** (Group 3)

**Group 2.** Figure 4 shows the efficiency achieved by the six indexes for the **EH** and **CH** configuration varying the number of pivots. **D-cache** again suffers from high internal complexity, reducing the linear scan time by 10% for 20 pivots at **EH**, even though it can save more than 50% of distance evaluations. Both **LaesaR** and **LaesaQ** starts slightly better than linear scan, but then their internal complexity dominates the search-time. This behavior might be due to the high intrinsic dimensionality that **EH** presents implying that any static selection of pivots will achieve bad performance. However, because **EH** fits a snake distribution, with just a few dynamic pivots **D-cache** and the Snake Table can discard more than 50% of distance evaluations. For a few pivots, **SnakeV3** achieves the best performance, however as the number pivots increases, the **SnakeV1** improves, mainly due to its lower internal complexity. As a summary for this experiment, the Snake Table achieves the best performance, enabling the indexing of metric spaces with high intrinsic dimensionality if they present a snake distribution.

**Group 3.** Figure 5 shows the efficiency achieved by the six indexes for the **ECK** and **EK3** configurations varying the number of pivots. In these configurations, **D-cache** starts to show better results, outperforming LAESA, because the saved distance computations pay for the internal complexity. However, with more than 10 pivots, **D-cache** begins to increase its search times. In this case, the exploitation of snake distribution becomes a remarkable approach for improving efficiency. While **LaesaQ** and **LaesaR** can discard 20% distances, the Snake Table can discard more than 50% of distances. In particular, **SnakeV3** (i.e., to store the last $p$ pivots by object) shows the fastest search times.

## 6   Conclusions and Future Work

In this work we presented the Snake Table, which achieves high performance for processing streams of queries with snake distribution. This satisfactory performance is due to its properties of dynamic selection of good pivots and low internal complexity. The Snake Table is able to reduce the search time for both fast and time-expensive distances, even in spaces with high intrinsic dimensionality. In particular, the Snake Table is a better alternative than D-cache in the tested scenarios.

The Snake Table presents an approach to index spaces when consecutive queries are similar among them. This behavior usually appears in content-based video retrieval (when the queries are consecutive keyframes), and it also may appears in interactive multimedia retrieval systems (when the user selects a new query object from the answers of a previous query). In a more general domain, given an unsorted set of queries, the test of snake distribution presented in this work may be useful to determine an optimal ordering of queries which will achieve a high performance in the Snake Table.

One usage of the Snake Table is to create an index for each stream of queries. When a user connects to the database, an empty Snake Table may be associated with the session. As the user performs queries with snake distribution, the Snake Table improves its performance because it will contain pivots close to the next queries. However, the Snake table is not memory efficient, because it requires space proportional to the size of the dataset and to the number of sessions connected. This approach is more suitable for medium-sized databases with long $k$-NN streams. Moreover, because it does not need to use a central shared index structure, it is also suitable for highly dynamic datasets.

On the one hand, pivots in a sliding window with snake distribution satisfy one desirable property: they should be close to either the query or the collection objects. On the other hand, those pivots do not satisfy other desirable property: they should be far away from each other. Hence, using a Snake Table with many pivots will only increase the internal complexity without increasing the efficiency because pivots will be mostly redundant. One approach to overcome this issue is to combine dynamic pivots with static pivots while resolving the stream. As it is shown in the experimental section, static pivots in the queries (**LaesaQ**) perform almost identically (sometimes even better) that static pivots in the reference objects (**LaesaR**). An improvement may be a combination between the SSS algorithm and the Snake Table. The Snake Table may chose to fix one of the dynamic pivots (i.e., to not remove it from the table) when it is far away from all the previous pivots, thus when the sliding window moves away, the fixed pivots will start to behave as static pivots complementary to the dynamic ones. Finally, LAESA can benefit of multi-core architectures by sharing the pivot table and resolving each query in different threads. However, it is not evident how to efficiently resolve parallel queries in the Snake Table due to the dynamic nature of its structure. Every thread should lock the pivot table to add the new distances, but this will interfere with other threads reading the table. A possible solution for this issue is to partition the queries into independent subsets, and each subset is resolved in an independent thread using its own Snake Table. We plan to address these issues in the future.

# References

1. Barrios, J.M., Bustos, B.: Competitive content-based video copy detection using global descriptors. Multimedia Tools and Applications, 1–36 (2011)
2. Bustos, B., Pedreira, O., Brisaboa, N.: A dynamic pivot selection technique for similarity search. In: Proc. of the Int. Workshop on Similarity Search and Applications (SISAP 2008), pp. 105–112. IEEE (2008)
3. Chávez, E., Navarro, G., Baeza-Yates, R., Marroquín, J.L.: Searching in metric spaces. ACM Computing Surveys 33(3), 273–321 (2001)
4. Kim, C., Vasudev, B.: Spatiotemporal sequence matching for efficient video copy detection. IEEE Transactions on Circuits and Systems for Video Technology 15(1), 127–132 (2005)
5. Law-To, J., Joly, A., Boujemaa, N.: MUSCLE-VCD-2007: a live benchmark for video copy detection (2007), http://www-rocq.inria.fr/imedia/civr-bench/
6. Manjunath, B.S., Ohm, J.-R., Vasudevan, V.V., Yamada, A.: Color and texture descriptors. IEEE Transactions on Circuits and Systems for Video Technology 11(6), 703–715 (2001)
7. Micó, L., Oncina, J.: A constant average time algorithm to allow insertions in the LAESA fast nearest neighbour search index. In: Proc. of the Int. Conf. on Pattern Recognition (ICPR 2010), pp. 3911–3914. IEEE (2010)
8. Micó, M.L., Oncina, J., Vidal, E.: A new version of the nearest-neighbour approximating and eliminating search algorithm (AESA) with linear preprocessing time and memory requirements. Pattern Recognition Letters 15(1), 9–17 (1994)
9. Skopal, T., Lokoč, J., Bustos, B.: D-cache: Universal distance cache for metric access methods. IEEE Transactions on Knowledge and Data Engineering 24(5), 868–881 (2012)
10. Vidal, E.: New formulation and improvements of the nearest-neighbour approximating and eliminating search algorithm (AESA). Pattern Recognition Letters 15(1), 1–7 (1994)
11. Zezula, P., Amato, G., Dohnal, V., Batko, M.: Similarity Search: The Metric Space Approach (Advances in Database Systems). Springer (2005)

# Algorithmic Exploration of Axiom Spaces for Efficient Similarity Search at Large Scale

Tomáš Skopal and Tomáš Bartoš

SIRET Research Group, Faculty of Mathematics and Physics,
Department of Software Engineering,
Charles University in Prague, Czech Republic
{skopal,bartos}@ksi.mff.cuni.cz
http://siret.cz

**Abstract.** Similarity search is becoming popular in even more disciplines, such as multimedia databases, bioinformatics, social networks, to name a few. The existing indexing techniques often assume the metric space model that could be too restrictive from the domain point of view. Hence, many modern applications that involve complex similarities do not use any indexing and use just sequential search, so they are applicable only to small databases. In this paper we revisit the assumptions which persist in the mainstream research of content-based retrieval. Leaving the traditional indexing paradigms such as the metric space model, our goal is to propose alternative methods for indexing that shall lead to high-performance similarity search. We introduce the design of the algorithmic framework **SIMDEX** for exploration of analytical properties (axioms) useful for indexing that hold in a given complex similarity space but were not discovered so far. Consequently, the known axioms will be localized as a subset within the universe of all axioms suitable for indexing. Speaking in a hyperbole, for database research the discovery of new axioms valid in some similarity space might have an impact comparable to the discovery of new laws of physics holding in parallel universes.

## 1 Introduction

For a long time, the database-oriented research of similarity search employed the definition of similarity restricted to the metric space model. Due to the fixed properties of *identity, positivity, symmetry*, and especially *triangle inequality*, metric distances enable to index a database for efficient querying using *metric indexes* [4,19,15], preventing thus from searching the whole database sequentially. Together with the increasing complexity of data types across various domains, recently there also appeared many *nonmetric* similarity functions [17].

Nowadays, we identify two types of research groups concerned with different aspects of similarity search – *database experts* and *domain experts*. The database experts deal with performance issues of similarity search and (mostly) do not care of particular domain applications. They just assume a similarity model that is constrained by some properties useful for database indexing, such as the metric space axioms. On the other hand, the domain experts (e.g., computational biologists) model similarities for specific practically oriented applications, while

G. Navarro and V. Pestov (Eds.): SISAP 2012, LNCS 7404, pp. 40–53, 2012.

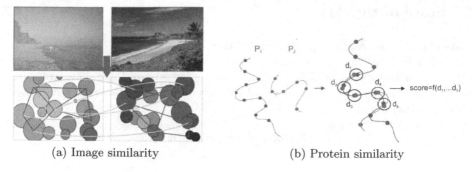

| (a) Image similarity | (b) Protein similarity |

**Fig. 1.** Sample similarity models

they do not (like to) care of any database-specific requirements, as this is outside their job area. As the result, only the simplest similarity models comply with the objectives of both groups. For example, various histogram representations of image features are largely measured using the Euclidean distance (or generally metric Minkowski $L_p$ distances, $p \geq 1$).

Besides simple models, domain experts often develop similarity models involving more sophisticated features or complex similarity functions. Such models better reflect the desired concept of similarities and lead to more effective/precise retrieval (see Fig. 1a for a sketch of robust matching using local image features). Naturally, the more complex similarity function the domain experts come with, the lower the likelihood is that it will be a metric distance. The problem of "nonmetricity" is not limited just to complex algorithms because even a slight change of a well-known metric distance could lead to a nonmetric, e.g., $L_p$ distances ($0 < p < 1$) that are used for robust matching of histograms [9]. More complex nonmetric similarities include various alignment algorithms for measuring functional similarity of protein sequences [18] or structures [6] (see Fig. 1b).

In summary, because of the not really integrated research efforts, both groups (database experts and domain experts) head into trouble in the near future.

- **Database research** – Current efficient solutions for constrained similarity models (e.g., based on the metric space model) might not be applicable to the future state-of-the-art similarity search problems. Simply, the database technology might provide only solutions for trivial or obsolete models.

- **Research in various domains** – assumes the sequential search as sufficient. As the models become so complex and/or the databases become so large-scale, an efficient database solution would become a critical requirement.

Based on this negative perspective, the major challenge and the main goal of our research is to find a complex solution that provides the various domain experts with database techniques that speedup similarity search yet that do not require any database-specific intervention to the similarity models. Before we outline our idea and the ground-breaking nature of the proposed algorithmic framework (see Section 5), we shortly summarize the state of the art and the previous attempts to unconstrained similarity search.

## 2    State of the Art

Instead of expensively computing all distances $\delta(q, o_i)$ to filter out database objects $o_i$ irrelevant to a given query $q$, the so-called *metric indexes* [4,19,15] use cheap lowerbound distances $LB(\delta(q, o_i))$. These lowerbound distances are computed using the triangle inequality and the precomputed distance to a reference object (*pivot*) $p$ as

$$LB_\triangle(\delta(q, o_i)) = |\delta(q, p) - \delta(p, o_i)|. \tag{1}$$

Naturally, having multiple pivots $p_j$, the particular lowerbounds (Eq. 1) could be combined in order to obtain tighter or better distance approximation. When using the *pivot table* [12] as the metric index, the lowerbound inequalities can be used directly, while other metric indexes require additional aggregation, metric region construction, etc. In Fig. 2a we can see the lowerbound distance and an object $o$ outside the query ball being filtered out from search without actually needing to compute $\delta(q, o)$.

As we mentioned earlier, there constantly appear *nonmetric* (unconstrained, respectively) similarity models [17]. The trend towards nonmetric models is not a marginal experience but rather a rule because effective applications usually require complex unconstrained similarity models. The response of the database research to this trend is, however, inappropriate. Almost all of the general-purpose database techniques that were designed to support nonmetric similarity models map the problem to the metric one and use metric or spatial access methods. In particular, the *TriGen algorithm* [16] applies a system of concave functions to the nonmetric distance to obtain an approximately metric behavior.

Older approaches directly map the data into some $L_p$ space while it suffers from an unpleasant trade-off. If the mapping preserves the similarity orderings, i.e., provides query results exactly the same as the nonmetric version, then it usually suffers from high intrinsic dimensionality [4]. Or, the mapping is only approximate and the search is fast but simultaneously introducing a retrieval error. In some cases, as in multimedia retrieval, the trade-off could provide satisfactory results. In many other situations, however, any loss in retrieval precision is unacceptable. For example, in biological, medical, or biometric applications every percent of precision counts heavily. For these critical applications the database research has no general solution so far.

## 3    Research Motivation

The motivation for our framework comes from an alternative approach to similarity indexing. Instead of "forcing" the distance and/or data to comply with the metric space model, for some data spaces it could be more advantageous to employ completely different indexing model that provides cheap construction of lowerbounds. The recently introduced ptolemaic indexing [7] could serve as an example for this conceptual shift.

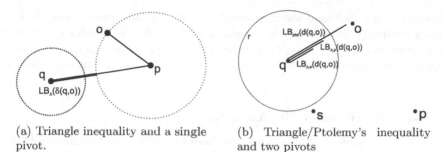

(a) Triangle inequality and a single pivot.

(b) Triangle/Ptolemy's inequality and two pivots

**Fig. 2.** Computing lowerbound distances using different approaches

## 3.1   Ptolemaic Indexing

In metric lowerbounding, the triangle inequality is used to construct lowerbounds for the distance. Analogously, in *Ptolemaic indexing* [7,11], *Ptolemy's inequality* is used to construct such lowerbounds. A distance function is called a *Ptolemaic distance* if it holds the properties of *identity*, *positivity*, *symmetry*, and satisfies *Ptolemy's inequality*. If a Ptolemaic distance also satisfies the *triangle inequality*, it is a *Ptolemaic metric*.

Ptolemy's inequality states that for any quadrilateral, the pairwise products of opposing sides sum to more than the product of the diagonals. In other words, for any four database objects $x$, $y$, $u$, $v \in \mathcal{D}$, we have the following:

$$\delta(x,v) \cdot \delta(y,u) \leq \delta(x,y) \cdot \delta(u,v) + \delta(x,u) \cdot \delta(y,v) \tag{2}$$

One of the ways the inequality can be used for indexing is in constructing the pivot-based lower bound. For a query $q$, object $o$, and pivots $p$ and $s$, we get the *candidate bound*:

$$\delta_C(q,o,p,s) = \frac{|\delta(q,p) \cdot \delta(o,s) - \delta(q,s) \cdot \delta(o,p)|}{\delta(p,s)} \tag{3}$$

For simplicity, we let $\delta_C(q,o,p,s) = 0$ if $\delta(p,s) = 0$. As for triangular lowerbounding, one would normally have a set of pivots $\mathbb{P}$, and the bound can then be maximized over all (ordered) pairs of distinct pivots drawn from this set, giving us the final Ptolemaic bound [7,11]:

$$\delta(q,o) \geq \mathrm{LB}_{\mathrm{ptol}}(\delta(q,o)) = \max_{p,s \in \mathbb{P}} \delta_C(q,o,p,s) \tag{4}$$

As for the metric case, the Ptolemaic lowerbound could be used to filter objects $o_i$ not contained in the query ball with radius $r$: $\mathrm{LB}_{\mathrm{ptol}}(\delta(q,o_i)) > r$.

In Fig. 2b we can see a real example (in two-dimensional Euclidean space) which shows that ptolemaic lowerbounding could provide tighter distance estimates than the metric one. Any of the lowerbounds constructed using the triangle inequality and two pivots $p$, $s$ would not filter the object $o$, as the value is lower than a radius of the range query. On the other hand, the ptolemaic lowerbound leads to a better distance approximation, and so the object $o$ is filtered out.

The ptolemaic indexing was successfully used with *signature quadratic form distance* [11] that was proved to be ptolemaic metric which is suitable for effective matching of image signatures [2]. In experiments the ptolemaic indexing applied to pivot table outperformed the corresponding metric variant four times in real-time cost.

## 3.2   Can We Automate the Axiom Exploration?

The idea of ptolemaic indexing shows that finding new axioms suitable for indexing could be a solution to speeding up (exact) similarity search in other way than mapping the problem to the metric space model. On the other hand, doing so *manually* would be even harder than forcing a domain expert to implant the metric axioms into her/his nonmetric distance. Note that even proving the "simple" triangle inequality axiom in the "simple" Euclidean distance is quite a complicated task, let alone proving the complex Ptolemy's inequality in a nonmetric distance implemented by a complex heuristic algorithm. Hence, the challenging question is whether we can automate the process of axiom exploration.

## 4   Framework Objectives and Main Goals

The main goal of our work is to develop an algorithmic framework for automatic exploration of axiom spaces for efficient similarity search at large scale. This framework, we call it **SIMDEX**, offers a complex solution that provides the various domain experts with database techniques that speedup similarity search yet do not require any database-specific intervention to the similarity models.

As the main input, we consider a particular similarity space described by (1) a *black-box distance function* and (2) a *database sample*. The "mining field" is a set of distances (or the corresponding distance matrix) obtained by computing the distances between pairs of objects in the database sample.

The basic idea of our framework is very straightforward – an iteratively constructed universe of expressions in their lowerbound forms is tested against the distance matrix by multiple evaluations. If all interpretations pass the test, the expression is declared as an *axiom* valid in the given similarity model.

Using this simple idea, we are able to algorithmically explore axiom spaces specified in a syntactic way – that is, we are not using a single canonized form and a tuning parameter (as the *TriGen* or other mapping approaches do). Consequently, the lowerbounding forms of triangle inequality (Eq. 1) and Ptolemy's inequality (Eq. 3) could be rediscovered as two instances in the axiom universe. Since the resulting set of analytical properties (axioms) will be obtained in the form of lowerbounding filters, it can be immediately used for indexing in the pivot table the same way as ptolemaic indexing was implemented [11].

Note that empirical testing of the expressions using multiple interpretations of $\delta(x, y)$ on the distance matrix leads just to the empirical evidence that an expression holds for the given model. However, this cannot be replaced by an analytical resolution that absolutely confirms the evidence, although in our case, a large number of positive tests could be treated as a sufficient confirmation.

## 4.1   Breaking the Metric Paradigm

The idea of our framework is based on a ground-breaking fundamental research – at least within the scope of data engineering and database systems. To the best of our knowledge, there has never been proposed such a complex framework for mining properties from similarity data as in this case. Although there have been many proposals introducing alternative perspectives on data indexing and mining, they were always based on a single mathematical model (such as the metric space model, ptolemaic indexing, fuzzy logic, multidimensional scaling, neural networks, etc.). In our framework, we do not a priori select a particular mathematical foundation, as we analytically discover the foundation itself.

Hence, the impact of the proposed framework is two-fold. From the technical point of view, the discovery of new axioms will enable large-scale similarity search in many applications where content-based retrieval is the essential component, e.g., multimedia retrieval, time series databases, biometric databases, etc. Since applications come from different domains outside the computer science, the contribution of our outcomes is truly multi-disciplinary.

From the philosophical point of view, the newly discovered expressions (or axioms) will contribute to the theoretic foundations of data engineering, data mining and disciplines beyond, such as computational geometry, geometric topology, and related disciplines. If a discovered axiom is general enough, it could open new horizons or research interests in many disciplines related to data engineering, similarity search, data mining, etc. and so the framework exhibits substantial inter-disciplinary nature.

# 5   Framework Methodology

In this section we formalize and describe into more details the methodology of the proposed algorithmic framework **SIMDEX**. We outline general requirements together with specific options applicable to our prototyped implementation of the framework. Note that we give the step-by-step tutorial of how to create and use the framework in different environments as the methodology is generally applicable regardless of the selected programming language or platform because it forms the theoretical basis and foundation. Furthermore, the framework might be modified, customized, or extended for specific purposes if necessary.

As we mentioned in the previous section, the input for the framework consists of a database sample $S \subseteq \mathcal{D}$, $|S| = n$, and a black-box distance function $\delta$. We want to obtain a set of expressions (*axioms*) $\mathcal{E}_\chi$ that are valid in the given model.

In order to proceed with methodology and the framework structure, we need to properly define the commonly used terms *expression* and *axiom*. Other terms and symbols that we use in the text are described in Table 1.

**Definition 1.** *An expression $E_i$ (or a candidate expression $E_i^*$) is a boolean expression that will be evaluated for any selection of $k$ objects from a database sample $S$:*

$$E_i : S^k \longrightarrow \{TRUE, FALSE\} \tag{5}$$

**Table 1.** Basic notation

| Symbol | Description |
|---|---|
| $\mathcal{D}, S \subseteq \mathcal{D}$ | database $\mathcal{D}$, a sample of database $S$ |
| $\delta(x, y)$ | distance between objects $x$ and $y$ ($\delta : \mathcal{D} \times \mathcal{D} \to \mathbb{R}$) |
| $M_{\delta,S}$ | distance matrix for sample $S$ and $\delta$ <br> $M_{ij} = \delta(o_i, o_j)$ where $o_i, o_j \in S$ |
| $M$ | the universe of distance matrices $M_{\delta,S}$ |
| $E_i, E_i^*, \mathcal{E}$ | expression $E_i$, candidate expression $E_i^*$, set of all expressions $\mathcal{E}$ |
| $A_i, \mathcal{A}$ | axiom $A_i$, set of all axioms $\mathcal{A}$ valid for a given model |

where $S^k = \overbrace{S \times S \times \ldots \times S}^{k}$ stands for k-tuple of different objects from S and the number k determines the cardinality of the expression $E_i$ which depends on the number of different variables used in the expression. The inequality always includes at least two objects from the sample database (query object q and database object o), therefore $k \geq 2$.

For clarity, we require a standardized lowerbound form of the expression $E_i$:

$$E_i \equiv \left[ \delta(q, o) \geq LB \right] \qquad (6)$$

where $\delta(q, o)$ is the real distance between the given query object q and a database object o, and the right-hand side (LB) is a non-terminal which is additionally expanded (see Section 5.2).

**Definition 2.** If an expression $E_i$ with cardinality k is TRUE for any k-tuple objects from the database sample S, we say that $E_i$ is an axiom ($A_i$).

More precisely, $A_i$ is an *empirical* axiom for the given model as it holds within the database sample S only and probably in $\mathcal{D}$. In order for $A_i$ to become a "real" axiom, it must be theoretically proved which is out-of-scope for this paper. Nevertheless, we will primarily focus on finding axioms.

Now and then, it might be useful to obtain expressions that are not always true, but are valid for majority of k-tuples, to enable fast but approximate values. For this purpose, we define the probability values $P_i$ (for expressions $E_i$).

**Definition 3.** The probability value $P_i$ for the given expression $E_i$ (with cardinality k) determines the ratio of positive occurrences of $E_i$ to all results:

$$P_i = Prob\Big( E_i(S^k) = TRUE \Big) = \frac{|E_i(S^k) \text{ returns } TRUE|}{|E_i(S^k) \text{ returns any value}|} \qquad (7)$$

The resulting value of $P_i$ depends purely on the number of times the expression $E_i$ is evaluated with a k-tuple. For optimization purposes, we could skip some of the $n^k$ tuples and get only the partial probability value $P_i^\bullet$ (but evaluated fast).

**Definition 4.** *The general algorithmic framework (SIMDEX) is defined as a tuple $\mathcal{F}(G, C, T)$ where $G$ is a grammar definition, $C$ is a set of conditions for generating expressions, and $T$ is a threshold probability value. Then, for a universe $M$ of distance matrices $M_{\delta,S}$, the framework acts as a function:*

$$\mathcal{F}(G, C, T) : M \longrightarrow \{(\mathcal{E}, \mathbb{R}_{[0,1]})\} \qquad (8)$$

*where $\mathcal{E}$ is a set of all expressions valid for a given model and $\mathbb{R}_{[0,1]}$ covers real numbers within the interval $[0, 1]$. The result is a set of pairs $(E_i, P_i)$ where each expression from the resulting set $E_i \in \mathcal{E}_X$ is valid for the sample $S$ with the corresponding probability $P_i$ ($\forall i : P_i \in [0,1], P_i \geq T$). The set of all resulting expressions $\mathcal{E}_X$ is a subset of $\mathcal{E}$: $\mathcal{E}_X = \{E_i\} \subseteq \mathcal{E}$.*

*If the threshold probability value $T = 1$, we can omit all probabilities $P_i$ and simplify the framework definition to*

$$\mathcal{F}(G, C) : M \longrightarrow \{\mathcal{A}\} \qquad (9)$$

*where $\mathcal{A}$ is a set of all axioms valid for the given model. Then, the result will be a set of axioms $\mathcal{E}_X = \{A_i\} \subseteq \mathcal{A}$.*

### 5.1 SIMDEX Framework Overview

In order to turn the framework into a practical and applicable tool, we divide the structure technically into several stages that need to be addressed as summarized in the text below. The complete overview of stages is depicted in Fig. 3.

- **(S1) Grammar Definition.** We create a grammar definition $G$ for generating expressions in the standardized form that meet specific requirements.

- **(S2) Expression Generation.** Because the grammar-based generation of expressions leads to an infinite universe, the set of tested inequalities needs to be optimized in order to discover the most promising expressions first.

- **(S3) Expression Testing.** Once a candidate expression is generated, it must be tested against the given matrix of distances $M_{\delta,S}$. Only such candidate expressions $E_i^*$ pass, for which their probability value $P_i$ in the given model is higher than the required threshold probability value $T$.

- **(S4\*) Expression Reduction.** As an optional step, we could investigate and apply heuristic techniques of expression combination and pruning.

- **(S5\*) Indexing Structures.** This step directly verifies the feasibility of the obtained indexing model according to the result set of expressions/axioms with an appropriate indexing structure (e.g., a pivot table).

- **(S6) Parallelization.** As the universe to explore is extensive, we demand parallelization to get satisfactory results within reasonable time period.

In the following sections, we describe all stages into more details and we better clarify and explain the framework definition. Also we outline the way of how we applied the general methodology when we created our prototype that covered stages S1–S3 and S6. The two remaining stages marked with an asterisk (S4\* and S5\*) are planned to be included in our prototype in the near future.

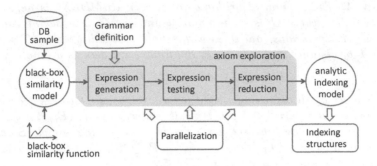

**Fig. 3.** SIMDEX Framework high-level schema

## 5.2   Grammar Definition (S1)

As the first step, we create or load a *grammar definition G*. We use the grammar theory specifically modified for the purpose of expression generation. Here, we need to define a grammar that will generate appropriate expressions $E_i$ that will become good candidates ($E_i^*$). The language defined by the grammar contains:

1. a reasonable number of *terminals*, such as
   (a) *object descriptor variables* ($q, o, p_1, p_2 \ldots$) and *constants* ($c_i$) where some of them are fixed and act as global reference points (such as pivots $p_j$); while others could stand for any object
   (b) *functions $f_j$* modifying not only the particular distances (as the triangle generating (*TG*) and triangle violating (*TV*) modifiers in TriGen algorithm [16]), but also whole expressions.
   (c) *standard arithmetic operators* ($+, *, -, /$), *numeric constants*, etc.

2. a limited number of their *combinations*

Speaking in precise terms, we refer but do not limit to *regular languages* or *L3/Type-3* grammar in Chomsky hierarchy [13] which is sufficient for our purposes. These languages are basically handled by a finite state automaton.

At the grammar level we need to guarantee that each generated expression $E_i$ is in the standardized form as outlined in Eq. 6 and would not be computationally too expensive compared to the direct distance computation between two objects $\delta(x, y)$. This means that the time complexity of computing a lowerbound value $O_{LB}$ is more effective than the one of the whole distance $O_\delta$: $O_{LB} \ll O_\delta$.

To mention other intuitive constraints, the expanded LB form cannot include $\delta(q, o)$, while it should include combinations of $\delta(q, p_j)$ and $\delta(p_j, o)$, where $p_j$ stands for a reference object for which the distances might be pre-computed.

## 5.3   Expression Generation (S2)

Even if the language and expansion recursion is limited, grammar-based generation is an exponential problem. Thus, the main objective is to guide the exploration to the most promising candidate expressions first, so that early termination of the exploration will end up with nonempty expression set. This discards inappropriate expression classes at early stage and saves time.

To achieve this, we will use a finite set of conditions for generating expressions $C = \{c_1, c_2, \ldots, c_m\}$ where each $c_i$ stands for a specific condition (e.g., discarding expressions $\frac{x}{x}$). Another specific criterion is *fingerprints* which are signatures of expressions that allow us to divide expressions into equivalent classes. For example, expressions $\delta(p_1, p_2)$ and $\delta(p_3, p_4)$, where $p_i$ is a variable, are equivalent for testing, so we test only one of them as the second one gives the same results.

To apply such conditions, we need a corresponding validating function that for a given expression $E_i$ returns TRUE to consider it further for testing or to refine it later, and FALSE otherwise. More precisely:

$$validate_C : \mathcal{E}_G \longrightarrow \{\text{TRUE}, \text{FALSE}\} \tag{10}$$

where $validate_C$ is a validation function for the given set of conditions $C$ and $\mathcal{E}_G$ is a set of expressions generated from the grammar $G$. The best situation is if the validating function reduces the set $\mathcal{E}_G$ to a large extent. At the same time, we try to maximize $\mathcal{E}_G \cap \mathcal{A}$, so we find all appropriate axioms. We mark the set of successfully validated expressions as $\mathcal{E}_\mathcal{V}^* = \{E_i | E_i \in \mathcal{E}_G : validate_C(E_i) = \text{TRUE}\}$ while $\mathcal{E}_\mathcal{V}^* = \{E_j^*\}_j \subseteq \mathcal{E}$. We want to minimize the set difference between sets $(\mathcal{E}_\mathcal{V}^* \setminus \mathcal{E})$ or ideally $(\mathcal{E}_\mathcal{V}^* \setminus \mathcal{A})$, so the number of candidates to be tested does not exceed too much all the expressions/axioms valid in the given model.

## 5.4   Expression Testing (S3)

Each validated candidate expression $E_i^* \in \mathcal{E}_\mathcal{V}^*$ is tested against the input matrix of distances $M_{\delta,S}$. Suppose the cardinality of the candidate expression $E_i^*$ is $k$ and $|S| = n$, we need to test the given expression for up to $n^k$ $k$-tuples. That is, all different terms in the inequality are interpreted multiple times using different sets of values from the matrix $M_{\delta,S}$. Based on variable names, the selection of distance values must be consistent (equally named variables get the same values).

Formally, we define the expression testing function $E\_test$ as

$$E\_test : (\mathcal{E}_\mathcal{V}^*, M) \longrightarrow \mathbb{R}_{[0,1]}$$

where $\mathcal{E}_\mathcal{V}^*$ is a set of candidate expressions to be tested, $M$ is the universe of input distance matrices $M_{\delta,S}$. The resulting real number determines the probability value $P_i$ with which the candidate expression holds in the specified model. If the probability value $P_i$ is greater than the threshold probability value $T$ ($P_i \geq T$), we mark the candidate expression $E_i^*$ as valid and add it to the result set of successfully tested expressions $\mathcal{E}_\mathcal{T}$. If $P_i = 1$, then the candidate expression $E_i^*$ is an axiom $A_i$.

In stages *Expression Generation* (S2) and *Expression Testing* (S3), where the hierarchy of expressions is generated and tested, we are utilizing combinatorial grammar coverage [10] and modern techniques from the constraint logic programming [1] that help with the reduction of the exponential size of the hierarchy using the standard expansion, inference, and pruning.

## 5.5   Expression Reduction (S4*)

To further condense the number of expressions we can refine the result by discarding weaker expressions, so only the best expressions/axioms will remain. For this, we define corresponding comparison criteria (such as applicability in the indexing structure; see Section 5.6).

Optionally (in this step) we try to combine the expressions $E_i^* \in \mathcal{E}_\mathcal{T}$ into a compound expression that would exhibit the tightest possible values for the smallest cost (the smallest size of the compound expression, number of reference objects used, etc.) We can obtain the compound expression $E^\circledast$ as

$$E^\circledast \equiv \left[ \delta(q,o) \geq \texttt{MaxLB}_{E_i \in \mathcal{E}_\mathcal{T}}(E_i) \right]$$

where $\mathcal{E}_\mathcal{T}$ is a set of successfully tested expressions and the function $\texttt{MaxLB}$ returns the right hand side of an expression $E_j$ which equals to expanded non-terminal LB that gives the maximal numeric value of the lower bounds for all $E_i$. Moreover, $\texttt{MaxLB}$ function might select different $E_j$ for different values of $\delta(q,o)$ which improves its filtering power but at the same time increases the complexity.

## 5.6   Indexing Structures (S5*)

This step verifies the resulting set of expressions/axioms $\mathcal{E}_\mathcal{X} = \mathcal{E}_\mathcal{T} \cup \{E_k^\circledast\}_k$ in practice within sample indexing tasks and validates the filtering power of each expression $E_i \in \mathcal{E}_\mathcal{X}$. Some of the existing indexing structures such as the pivot table [4,19] could be immediately used as an indexing structure for any kind of lowerbound expression $E_i$ that involves reference objects. As we require all expressions/axioms to be in the standardized form, they could be immediately included in the pivot table filtering options.

Nevertheless, it might happen that a particular expression/axiom enables the design of an advanced indexing structure based on the hierarchical decomposition of the database. Unlike pivot tables, such an index could also improve other costs than distance computations, such as I/Os or space complexity. The main benefit of direct applicability of obtained expressions within indexing structures is that we can easily and instantly use it to solve real-world problems and situations. This gives us even better feasibility study of the conducted research.

## 5.7   Parallelization (S6)

Despite the optimizations in stages S1–S4, we still have to assume the explored universe will be huge. Therefore, a massive parallelization of the exploration process must be taken into account from the very beginning. For this purpose, we consider and propose three parallel architectures to implement. For the preliminary experimentation we leverage (1) classic multi-core CPU systems with multi-threading. For future evaluations, we consider (2) vector parallelism based on multiple many-core GPU systems, (3) Map-Reduce technique [5] applied to a powerful CPU farm (ideally with the peer-to-peer distributed architecture), or to the supercomputer architecture with lots of cores.

---

**Algorithm 1.** SIMDEX $(G, C, T, S, \delta)$

---

**Require:** Grammar definition $G$, validation conditions $C$, threshold probability value $T$, database sample $S$, distance function $\delta$
1:   $M_{\delta,S} \leftarrow$ new distance matrix $(\delta, S)$
2:   $expressions \leftarrow$ ExpressionGeneration(G, C)
3:   **for all** $E_i$ in $expressions$ **do**
4:     **if** $validate_C(E_i)$ equals **false then**
5:       $expressions$.Remove($E_i$)   {validity check fails}
6:       continue   {skip further testing of the expression $E_i$}
7:     **end if**
8:     **if** $E\_test(E_i, M_{\delta,S}) < T$ **then**
9:       $expressions$.Remove($E_i$)   {probability test fails}
10:    **end if**
11: **end for**
12: **return** $expressions$   {remaining expressions compose the result set}

---

# 6 Experimental Evaluations

To validate the potential of the proposed theoretical algorithmic framework **SIMDEX**, we developed a prototype that covers stages S1–S3 and S6. We present an algorithm that gives an overview of the whole expression generating/validating/testing process (see Algorithm 1). All steps are straightforward and correspond to stages S1 through S3 as we presented in previous sections.

## 6.1 Experiments

We took five different datasets, each comprised of a distance matrix computed for 20 random database objects. For the grammar $G$, we used *standard arithmetic operators* $+, *, -, /$ together with three pivots.

We tested 25,000 expressions and for each expression we evaluated all possible variable assignments from the datasets' objects. During the tests, we looked for lowerbound tightness – min/max/avg difference between the real distance value compared to lowerbound value. The results are shown in Table 2 and for better transparency, we omit the left-hand sides, $\delta(q, o)$, of the resulting expressions and show only the expanded LB non-terminals. We evaluated these datasets:

- **Corel Image Features [8].** We used non-metric $L_p$ distance ($p = 0.5$).
- **CoPhIR [3].** For simplicity, we also used $L_p$ distance ($p = 0.5$).
- **Movie ratings.**[1] We took movie ratings as sets and used Jaccard coefficients to model similarities between them
- **Listeria.**[2] For exploration, we applied *Levenshtein (edit) distance.*
- **Spectrometry.** In this case, we applied the parameterized Hausdorff distance to the corresponding dataset [14].

---

[1] http://www.grouplens.org/node/73
[2] SISAP Metric Library, www.sisap.org

**Table 2.** Expression Evaluation results

| Dataset | Expression | Success Ratio | MIN | MAX | AVG |
|---|---|---|---|---|---|
| Corel | triangle inequality | 99 % | 0.0034 | 0.9983 | 0.3764 |
| | $\|\delta(q,p) \cdot \delta(o,p) \cdot (\delta(o,p) - \delta(q,p))\|$ | 100 % | 0.1059 | 0.9991 | 0.5020 |
| | $(\delta(q,p) - \delta(o,p))^2$ | 100 % | 0.1352 | 0.9999 | 0.5054 |
| | $\|(\delta(q,p_1) - \delta(o,p_1))(\delta(q,p_1) - \delta(o,p_2))\|$ | 100 % | 0.0420 | 0.9999 | 0.5161 |
| CoPhIR | triangle inequality | 97.5 % | 0.0021 | 0.9736 | 0.2696 |
| | $(\delta(q,p) - \delta(o,p))^2$ | 100 % | 0.0718 | 0.9979 | 0.3808 |
| | $\|(\delta(q,p_1) - \delta(o,p_1))(\delta(q,p_2) - \delta(o,p_2))\|$ | 100 % | 0.0845 | 0.9969 | 0.3935 |
| Ratings | triangle inequality | 100 % | 0.6067 | 1.0 | 0.9037 |
| | $\frac{1}{2 \cdot \delta(o,p)}$ | 100 % | 0.0119 | 0.5 | 0.4254 |
| | $(\delta(q,p_1) + \delta(o,p_1)) \cdot \frac{\delta(q,p_1)}{\delta(p_1,p_2)}$ | 100 % | 0.0103 | 0.5845 | 0.4254 |
| Listeria | triangle inequality | 99 % | 0 | 0.9559 | 0.1388 |
| | $\delta(p_1,p_2) \cdot \frac{1}{\delta(p_1,p_2) + \delta(o,p_2)}$ | 100 % | 0.0075 | 0.9994 | 0.2393 |
| | $\delta(q,p_1)^2 \cdot \frac{1}{\delta(o,p_2) \cdot \delta(q,p_2)}$ | 100 % | 0.0008 | 0.9985 | 0.2401 |
| | $(\delta(q,p_1) + \delta(o,p_1)) \frac{\delta(q,p_1)}{\delta(p_1,p_2)}$ | 100 % | 0.0032 | 0.9970 | 0.2555 |
| Spectrometry | triangle inequality | 100 % | 0.1823 | 0.93 | 0.7329 |
| | $\delta(o,p) - \delta(o,p)^2$ | 100 % | 0.0009 | 0.8758 | 0.6638 |
| | $\|(\delta(q,p_1) \cdot \delta(o,p_2)) - \delta(q,p_2)^2\|$ | 100 % | 0.0148 | 0.9399 | 0.7054 |

## 6.2  Summary

We explored several expressions for different datasets and found axioms that
might be directly used for indexing purposes. However, we still need to evaluate
the quality of the obtained expressions compared to existing models based on
triangle/ptolemaic inequalities. Then, we detect which axioms are applicable and
could become efficient variants for real-world situations. These steps remain as
our future work.

## 7  Conclusions

We proposed the algorithmic framework **SIMDEX** for indexing any similarity
model that leaves the restrictive paradigm of the metric space model and we
verified the feasibility of the framework through the implemented prototype. We
developed more generic approach to indexing by introducing this algorithmic
framework for exploration of axioms (analytical properties) that hold in a given
complex similarity space but were not discovered so far. Consequently, the previously known axioms will be localized as a subset within the universe of all
axioms suitable for indexing. The primary goal of **SIMDEX** framework is to
enable the complex similarity models to enter the competitive arena of industrial
practices in content-based retrieval that demand uncompromising performance.

In the future we plan to intensively elaborate and include all remaining stages
into our prototype while enhancing the existing ones. Our objective is to create
end-to-end framework implementation which will be suitable for any research
community and test it extensively for various different data models.

**Acknowledgments.** This research has been supported by Czech Science Foundation (GAČR) project 202/11/0968 and by Grant Agency of Charles University (GAUK) project 567312.

# References

1. Barták, R.: Constraint Models for Reasoning on Unification in Inductive Logic Programming. In: Dicheva, D., Dochev, D. (eds.) AIMSA 2010. LNCS, vol. 6304, pp. 101–110. Springer, Heidelberg (2010)
2. Beecks, C., Uysal, M.S., Seidl, T.: Signature quadratic form distance. In: Proc. ACM International Conference on Image and Video Retrieval, pp. 438–445 (2010)
3. Bolettieri, P., Esuli, A., Falchi, F., Lucchese, C., Perego, R., Piccioli, T., Rabitti, F.: CoPhIR: a Test Collection for Content-Based Image Retrieval. CoRR, abs/0905.4627v2 (2009)
4. Chávez, E., Navarro, G., Baeza-Yates, R., Marroquín, J.L.: Searching in metric spaces. ACM Computing Surveys 33(3), 273–321 (2001)
5. Dean, J., Ghemawat, S.: MapReduce: simplified data processing on large clusters. In: Proceedings of the 6th Conference on Symposium on Opearting Systems Design & Implementation - Volume 6 (2004)
6. Galgonek, J., Hoksza, D., Skopal, T.: SProt: sphere-based protein structure similarity algorithm. Proteome Science 9, 1–12 (2011)
7. Hetland, M.L.: Ptolemaic indexing. arXiv:0911.4384 [cs.DS] (2009)
8. Hettich, S., Bay, S.D.: The UCI KDD archive (1999), http://kdd.ics.uci.edu
9. Howarth, P., Rüger, S.: Fractional distance measures for content-based image retrieval. In: 27th European Conference on Information Retrieval, pp. 447–456. Springer (2005)
10. Lämmel, R., Schulte, W.: Controllable Combinatorial Coverage in Grammar-Based Testing. In: Uyar, M.Ü., Duale, A.Y., Fecko, M.A. (eds.) TestCom 2006. LNCS, vol. 3964, pp. 19–38. Springer, Heidelberg (2006)
11. Lokoč, J., Hetland, M.L., Skopal, T., Beecks, C.: Ptolemaic indexing of the signature quadratic form distance. In: Proceedings of the Fourth International Conference on Similarity Search and Applications, pp. 9–16. ACM (2011)
12. Navarro, G.: Analyzing metric space indexes: What for? In: IEEE SISAP 2009, pp. 3–10 (2009)
13. Noam, Chomsky: On certain formal properties of grammars. Information and Control 2(2), 137–167 (1959)
14. Novák, J., Skopal, T., Hoksza, D., Lokoč, J.: Non-metric Similarity Search of Tandem Mass Spectra Including Posttranslational Modifications. Journal of Discrete Algorithms 13 (2012)
15. Samet, H.: Foundations of Multidimensional and Metric Data Structures. Morgan Kaufmann Publishers Inc., San Francisco (2005)
16. Skopal, T.: Unified framework for fast exact and approximate search in dissimilarity spaces. ACM Transactions on Database Systems 32(4), 1–46 (2007)
17. Skopal, T., Bustos, B.: On nonmetric similarity search problems in complex domains. ACM Comput. Surv. 43, 34:1–34:50 (2011)
18. Smith, T.F., Waterman, M.S.: Identification of common molecular subsequences. Journal of Molecular Biology 147(1), 195–197 (1981)
19. Zezula, P., Amato, G., Dohnal, V., Batko, M.: Similarity Search: The Metric Space Approach (Advances in Database Systems). Springer-Verlag New York, Inc., Secaucus (2005)

# Polyphasic Metric Index: Reaching the Practical Limits of Proximity Searching

Eric Sadit Tellez, Edgar Chavez, and Karina Figueroa

Universidad Michoacana de San Nicolás de Hidalgo, México
sadit@lsc.fie.umich.mx,
{elchavez,karina}@fismat.umich.mx

**Abstract.** Some metric indexes, like the pivot based family, can natively trade space for query time. Other indexes may have a small memory footprint and still outperform the pivot based approach; but are unable to increase the memory usage to boost the query time. In this paper we propose a new metric indexing technique with an algorithmic mechanism to lift the performance of otherwise rigid metric indexes.

We selected the well known List of Clusters (LC) as the base data structure, obtaining an index which is orders of magnitude faster to build, with memory usage adaptable to the intrinsic dimension of the data, and faster at query time than the original LC. We also present a nearest neighbor algorithm, of independent interest, which is optimal in the sense that requires the same number of distance computations as a range query with the radius of the nearest neighbor.

We present exhaustive experimental evidence supporting our claims, for both synthetic and real world datasets.

## 1 Introduction

The metric indexing machinery can be used in diverse fields, such as pattern recognition, textual and multimedia information retrieval, machine learning, streaming compression, lossless and lossy compression, biometric identification and authentification, bioinformatics, among others [1]. However, proximity searching is a challenging problem since exact indexes (those returning exactly the objects contained in a range or nearest neighbor query, defined below) have a linear worst case on the size of the database, even when the query output set has $O(1)$ size. This behavior is thoroughly documented in the literature by Samet [2], Chavez et al. [3], Böhm et al. [4], Zezula et .al [5], and Pestov [6–8].

To cope with this intrinsic high dimensional case, the metric indexes should be tweaked to support approximate searches, as described in [9–11], and imply loosing some relevant answers to speed up the query time. Other relaxations include allowing reporting false positives, some examples are [12–18]. While the above relaxed approaches can be used in many application scenarios, the exact indexing problem is interesting by its own side.

In this work we introduce a new metric index very robust to the intrinsic dimension increase, with very good tradeoffs among memory, real searching time, and number

G. Navarro and V. Pestov (Eds.): SISAP 2012, LNCS 7404, pp. 54–69, 2012.

of computed distances. Moreover, our index can be engineered to allow the approximate and probabilistic relaxations cited above, broadening the usage spectrum of our technique.

Before discussing our framework in more detail, let us present some definitions and notation. A general metric space $(U, d)$ is composed by an universe of objects $U$, and a distance function $d : U \times U \rightarrow \Re$, such that for any $u, v, w \in U$, $d(u, v) > 0$ or $d(u, v) = 0 \iff u = v$, $d(u, v) = d(v, u)$, and $d(u, w) + d(w, v) \geq d(u, v)$. These properties are known as strict positiveness, symmetry, and the triangle inequality, respectively. The last property, is the main tool to filter candidates from a result set using the general metric space model.

Let $S$ be a database $S \subseteq U$, of size $n = |S|$, we considere two possible operations:

- *k nearest neighbor query.* Retrieve the $k$ closer elements of a query $q$ in $S$, formally $\text{k-nn}_d(q, S) = \{u \mid d(u, q) \leq d(v, q) \; \forall u, v \in S\}$ where $|\text{k-nn}_d(q, S)| = k$, or simply $\text{k-nn}(q, S)$ if the context provides enough information to avoid confusion.
- *range query.* Searching all objects around $q$ within a range $r$. It is defined as $(q, r)_d = \{u \in S \mid d(q, u) \leq r\}$, or simply $(q, r)$ if it is clear in the context.

There exists two main families of indexes tackling the proximity search problem using only information obtained by precomputed distances: pivot based indexes and compact partition indexes.

## 1.1 Pivot Index

The filtering with a set of pivots can be regarded as a contractive mapping from the original space $U$ to the vector space where the coordinates are the distances to the pivots. In other words, if $P = \{p_1, p_2, \cdots, p_m\} \subseteq U$ is the set of pivots, for $u, v \in S$ we define $D(u, v) = \max_{1 \leq i \leq m} |d(u, p_i) - d(v, p_i)|$. Using the triangle inequality, it is clear that $D(u, v) \leq d(u, v)$ and hence it is also clear that $(q, r)_d \subseteq (q, r)_D$.

## 1.2 Compact Partitions

A compact partition index creates regions where items are spatially close to each other. A tree based index selects a set of centers per node $c_1, c_2, \cdots, c_m \in S$, such that $c_i$ is the center of the subtree $T_i$. The set of centers induces a partition of the dataset such that each $T_i$ is spatially compact; for example, $u \in T_i$ if $d(c_i, u) = \arg\min_{1 \leq j \leq m} d(c_j, u)$. The covering radius $\text{cov}(c_i) = \max_{u \in T_i} d(c_i, v)$ is stored for each node. This construction is applied recursively. A query $(q, r)_d$ is solved recursively starting from the root node. If $d(q, c_i) \leq r$ then $c_i \in (q, r)_d$, and $T_i$ must be explored if $|d(q, c_i) - \text{cov}(c_i)| \leq r$.

**The List of Clusters.** The List of Clusters (LC) is a surprising data structure. With a fixed memory usage the LC outperforms all other indexes computing a smaller number of distances per query, specially when the data is high dimensional. A drawback of the LC is the high cost of construction, requiring a quadratic number of distances. This cost has the same origin of its unmatched performance.

Let us explore with some detail the construction and the searching algorithm for the LC, as described by Chávez and Navarro [19]. Define $I_{S,c,\mathsf{cov}(c)} = \{u \in S \setminus \{c\} \mid d(c, u) \le \mathsf{cov}(c)\}$ as the bucket of *internal* elements, which lie inside center ball of $c$, and $E_{S,c,\mathsf{cov}(c)} = \{u \in S \mid d(c, u) > \mathsf{cov}(c)\}$ as the rest of the elements (the *external* ones). Now the process is repeated recursively inside $E$. The construction procedure returns a list of triples $(c_i, r_i, I_i)$ (center,radius,bucket).

Please note that the number of centers is unknown beforehand. There are two possible parameters, the number of objects inside a ball and the radius of the ball. This intrinsically defines the number of centers. As suggested in the original paper we select the number of centers $m$, by selecting instead $n/m$ and avoid a complex parametrization of the algorithm.

When the intrinsic dimensionality of the data is high, then most of the clusters need to be reviewed. In [19] the authors used probabilistic arguments to bound the complexity of the searching to $O(n^\alpha)$ distance computations, for some $\alpha \le 1$ which depend on the intrinsic dimension of the data.

### 1.3 Other Composite Indexes

There exists some indexes in the literature being composed of several other indexes. Below, we list some methods using this scheme.

Gionis et al. [17] introduces a new approximate technique for similarity searching in high dimensional datasets. It is called as Locality Sensitive Hashing (LSH). These indexes are typically organized as hashing tables, such that objects in the same bucket are closer under a distance function $d$ with high probability [18, 17]. LSH offers distance based probabilistic guarantees, but the recall is low as compared with other approaches. Also, there exist a limit on the guarantees that can be ensured with a sigle instance of LSH. So, if results with higher quality are required as set of LSH instances should be used.

Kyselak et al. [20] observed in the literature that approximate methods optimize the average accuracy of the indexes. However, it is common to find very bad performances on individual queries, and consequently, an excel performance on others. So, they propose a simple solution to stabilize the accuracy. The idea is to reduce the probability of a poor-quality result using multiple independent indexes solving the same query, such that at least one index achieves good quality on a result.

One important point, to put our work in perspective, is to notice that in the literature, assembling multiple indexes happen only in the context of approximate techniques. The focus is on increasing the recall (or the quality of the result, in general). And that comes at the price of increase both searching and preprocessing time. In contrast our solution *decrease both* searching and preprocessing times.

In a slightly related paper [21], author Yianilos introduced the Vantage Point Forest (VPF). Here the idea is to have a single index composed by a collection of trees. The idea is to exclude in each tree those objects near the frontier of the left and right branches, at every node. Each tree will index only a portion of the database. In exchange each tree will avoid backtracking, hence speeding up the search. There is a delicate balance between backtracking and searching sequentially in a collection of trees. Additionally, the searching radius to avoid backtracking is way too small for most applications.

In paper [22], Skopal introduced the PM-Tree. This is a peculiar index mixing pivot based and compact partition indexes. The PM-Tree base-algorithm is the secondary memory index M-Tree (Ciaccia et al. [23]) enriched with global scope pivots on top of the local compact partitions of the M-Tree. We will show that our approach can mimic this mixed type of indexes, while reducing the complexity and being less cumbersome than the PM-tree.

### 1.4   Our Contribution

A basic view of a metric index is to regard it as a partition of the space. The index then guide the search by filtering some parts for each particular query. The parts not filtered are then exhaustively checked. Our algorithmic idea is to use several indexes, several partitions, applying the corresponding filters and then search in the intersection of all the non-filtered parts. One key aspect of the above idea is to efficiently implement union/intersection operations to quickly obtain the answer.

We propose novel algorithms for proximity searching, based on fast union-intersection operations. Specifically, we introduce algorithms to solve range and k-nn searches. Our algorithm is optimal in the same sense given by Samet et al [2], where the necessary number of computed distances to solve k-nn queries is the same than a range search with the proper searching radius.

Since our index is composed of several underlaying indexes one requirement should be to build on the better brick. We have to select the better index, appropriate for intrinsically high dimensional data. Unfortunately the options are scarce. The most robust indexes are expensive either in memory usage or preprocessing time, as detailed below:

- AESA [24] stores $O(n^2)$ distances, and the cost of construction is of the same order. Moreover, it requires a quadratic number of arithmetic and logical operations at query time. However it requires (experimentally, on average) a fixed number of distance computations to solve a query, on a dataset with fixed intrinsic dimension.
- The list of clusters (LC) [19] uses $O(n)$ integers for the index, and $O(m)$ distances, with $m$ the number of centers. The construction cost is roughly $mn/2$ distance computations, $O(mn)$. However, as explained by Chavez and Navarro [19], high intrinsic dimensional datasets require $n/m = O(1)$ to be useful, implying $O(n^2)$ distances on the preprocessing step.

The above costs are even worst for our case, since we need many indexes to be built and queried. In practical terms this restrict us to low-cost indexes, in both space and preprocessing time. Neither AESA nor the LC (for high dimensional datasets) are suitable (as is) for the task for the prohibitive construction time and/or storage costs.

To be able to use the LC, keeping the construction time bounded, we used a suboptimal selection of the governing parameter $m/n$ in each one of the indexes. This implies a not so large number of centers, as $m = O(n^\beta)$ for $\beta < 1$. We will use $\lambda$ randomized instances of the LC to boost the filtering. We bet on the fact that the probability of being discarded by at least one index increases with the number of indexes. The price to pay is an increase on the storage in a small factor $\lambda$. We will provide a probabilistic model of the search performance of our composite index and will verify it experimentally.

Summarizing, we obtain a powerful metric index with $O(\lambda n^{1+\beta})$ preprocessing time, $O(n^\alpha)$ search time. With $\alpha, \beta < 1$ and $\lambda$ is a small integer number. Also, our index requires $O(\lambda n)$ identifiers. Our algorithmic proposal is general enough to support any mixture of indexes, beyond our proposed modifications of the LC. Furthermore, the discarding rule is arbitrary and this imply it may not be based of the triangle inequality. This sole feature have not been proposed before in the literature, up to the best of our knowledge.

## 2   The Polyphasic Metric Index (PMI)

Let $\Lambda = \{T_i\}$ be a collection of metric indexes, $\lambda = |\Lambda|$. Each $T \in \Lambda$ induces a partition $\Pi_T$ of the database, as it is standard for metric indexes. $L_j \in \Pi_T$ denotes the $j$-th part of $\Pi_T$. Most metric indexes fall on this categorization, since all of them are based on equivalence classes as described by Chavez et al. [3]. Some indexes have complex decision rules and it is difficult to take advantage of the implicit partition. One notable exception is the LC. The implicit partition of the DB coincides with an explicit partition very easy to handle. Each bucket $I_{c_i}$ is a part. Moreover, the set of centers $C$ in the LC is itself a part (see section 1.2).

Below we describe with detail the basic algorithms for the generic index.

## 3   Range Search

For a query $(q, r)_d$, define $C_\Lambda$ the set of elements not filtered by any of the indexes in $\Lambda$. To avoid cumbersome notation, we do not include the explicit dependence on $(q, r)_d$ in $C_\Lambda$.

Solving $(q, r)_d$ imply computing the distance between the query and every element in $C_\Lambda$ and reporting only those with distance less than $r$ to the query.

$C_\Lambda$ is the intersection of all $C_T$, i.e. the candidate set for $T \in \Lambda$. $C_T$ is computed retrieving all lists not being discarded by the triangle inequality, and then joining them. Algorithmically speaking, the range search is a set union-intersection algorithm. Formally, we must compute $C_T = \bigcup_{L \in \mathcal{L}_{T,(q,r)}} L$. Where $\mathcal{L}_{T,(q,r)}$ is the set of all parts in $\Pi_T$ such that the discarding rule of the incumbent index cannot filter. To fix ideas think in this rule as the triangle inequality. Finally, the complete candidate list is computed as $C_\Lambda = \bigcap_{T \in \Lambda} C_T$. Notice that the intersection $C_\Lambda$ is the set of objects not filtered by the discarding rules (triangle inequality) using all indexes in $\Lambda$.

Please notice that a center can be shared by some backend indexes, specially when $m$ is large, such that we might be duplicating distance evaluations against some centers. We can select centers to be disjoint at construction time. Our implementation uses a simpler solution, we add a cache of distances per query $q$ such that a $d(q, c_i)$ is evaluated once in the solving process. The same strategy would be applied to nn searching (see section 4).

There are several optimal (on the size of the sets being intersected) union and intersection algorithms using sorted lists as abstractions of sets. Some examples are presented by Hwang and Lin [25], and Baeza-Yates [26]. Furthermore, beyond the optimality on the size, there are several instance optimal algorithms, like those presented by Demaine et al. [27], and Barbay and Kenyon [28]. Since our parts are of fixed size (we are using LC to partition the dataset), and we cannot ensure *easy instances* for instance optimal algorithms, we focus on producing a fast worst case union-intersection algorithm not in the comparison model. Additionally, the structure of our algorithm is used to solve the k-nn searching with dynamic programing, producing the algorithm of the next section.

The algorithm 1 shows a fast $\Theta(\sum_{T \in \Lambda} |\mathcal{C}_T|)$ union and intersection algorithm. In the algorithm, the array $A$ is explicitly stored because $n$ is not so large in practice (a few millions at most). If the plain storage of $A$ in memory is not feasible, it can be easily replaced by a hash table, the complexity holds on average.

---

**Algorithm 1.** Union-intersection algorithm

---

**Input:** $\mathcal{L}_{T,(q,r)}$ for all $T \in \Lambda$.
**Output:** The candidate set $\mathcal{C}_\Lambda$, i.e., the set of buckets that cannot be discarded using the triangle inequality.

1: Let $A[1, n]$ be an array of integers initialized to zero, each item has $\lceil \log(\lambda - 1) \rceil$ bits
2: Let $\mathcal{C}_\Lambda = \emptyset$
3: **for** $T \in \Lambda$ **do**
4:     **for** $L \in \mathcal{L}_{T,(q,r)}$ **do**
5:         **for** $u \in L$ **do**
6:             **if** $A[u] + 1 = \lambda$ **then**
7:                 $\mathcal{C}_\Lambda \leftarrow \mathcal{C}_\Lambda \cup \{u\}$
8:             **else**
9:                 $A[u] \leftarrow A[u] + 1$
10:             **end if**
11:         **end for**
12:     **end for**
13: **end for**

---

## 4 Nearest Neighbor Search

Even when we focus on the nearest neighbor search, the procedure is trivially extended to solve k-nn queries. Our nearest neighbor search algorithm is based on the *best first* strategy described by Samet [2] and our union-intersection algorithm (alg. 1).

The nearest neighbor searching is solved in algorithm 2. In this procedure there are three special variables, $r_\top^*$, $r_\bot^*$ and $q^*$. $r_\top^*$ is the best upper bound of the covering radius for our query at any moment, $r_\bot^*$ the best known lower bound, and $q^*$ is the best known candidate to be nn(q). At the beginning, $r_\top^* = \infty$, $r_\bot^* = 0$, and $q^* = $ undefined; at the end of the procedure, $r_\top^* = r_\bot^* = d(q, \text{nn}(q))$,[1] $d(q, q^*) = d(q, \text{nn}(q))$, and $q^*$ is nn(q).

---

[1] Ideally, $r_\bot^*$ will stop in $r_\top^*$, but it could overrun $r_\top^*$ since it advance in ranges.

---

**Algorithm 2.** Best first nearest neighbor search

---

**Input:** A query object $q$.
**Output:** $r_\top^* = d(q, \mathsf{nn}(q))$ and $q^* = \mathsf{nn}(q)$.

1: Let $A[1, n]$ be an array of integers initialized to zero, each item has $\lceil \log (\lambda - 1) \rceil$ bits.
2: Let $q^*$ be the best candidate at any moment of the nearest neighbor, $q^* \leftarrow$ undefined.
3: Let $r_\bot^* = 0$
4: Let $r_\top^*$ be the best guest at any moment of $d(q, \mathsf{nn}(q))$, $r_\top^* \leftarrow \infty$.
5: **while** $r_\bot^* \leq r_\top^*$ **do**
6:     advance_bottom $\leftarrow$ true
7:     **for** $T \in \Lambda$ **do**
8:         {Inside next_best both $r_\top^*$ and $q^*$ should be adjusted if it is necessary.}
9:         $L \leftarrow$ next_best$(T)$
10:         **for** $u \in L$ **do**
11:            advance_bottom $\leftarrow$ false
12:            **if** $A[u] + 1 = \lambda$ **then**
13:              **if** $d(q, u) \leq r_\top^*$ **then**
14:                $r_\top^* \leftarrow d(q, u)$
15:                $q^* \leftarrow u$
16:              **end if**
17:            **else**
18:              $A[u] \leftarrow A[u] + 1$
19:            **end if**
20:         **end for**
21:         **if** advance_bottom **then**
22:            Increase $r_\bot^*$ to the minimum radius such that at least another candidate (in any $T \in \Lambda$) list will be available.
23:         **end if**
24:     **end for**
25: **end while**

---

The objective is to convert the nn search in a sequence of range searches. In each internal range search the covering radius $r_\top^*$ can be reduced, while $r_\bot^*$ is increased. The algorithms follows the constraint $r_\bot^* \leq r_\top^*$. The main problem is that we require several times to perform union-intersection operations over the same parts, as required by the algorithm of range search. More detailed, let $r_\bot^*$ be decomposed in its steps in the algorithm, thus let $r_\bot^{*h}$ be $h$-th value of $r_\bot^*$ at the $h$ step. Since $r_\bot^{*1} \leq r_\bot^{*2} \leq \cdots \leq r_\bot^{*s}$, after $s$ steps, then follows that $(q, r_\bot^{*1})_d \subseteq (q, r_\bot^{*2})_d \subseteq \cdots \subseteq (q, r_\bot^{*s})_d$. Fortunately, range searches can be decomposed in terms of the previous ones, hence we can use our union-intersection (algorithm 1) since it stores in $A$ the cardinality of the intersection of previous steps.

Let us define next_best$(T)$ as the procedure returning at each call a list *not yet visited*, such that this list intersects the current query ball, i.e. $(q, r_\bot^*)$. It is necessary to remark that next_best$(T)$ can adjust $r_\top^*$ and $q^*$ as needed. The idea behind next_best$(T)$ is to access each $L \in \Pi_T$ in the order that it should be accessed by consecutive range searches $(q, r_\bot^*)$. In each step $r_\bot^*$ is increased (line 22 in algorithm 2) to the minimum necessary to obtain another $L$.

---

**Algorithm 3.** Global view of the next_best($T$) procedure

---

**Initialize:** Let $\mathcal{L}_{T,(q,r_\perp^*)}$ be the set of lists intersecting the minimum radius ball containing $(q, r_\perp^*)$.

**Input:** Let $r_\perp^* \leftarrow 0$.

**Output:** A set (list) containing objects intersecting the current query ball.

**Procedure:** At each call it proceeds as follows:

1: **if** $r_\perp^* > r_\top^*$ **then**
2:     **return** $\emptyset$
3: **else**
4:     **if** $r_\perp^*$ was incremented **then**
5:         Retrieve the necessary lists to complete $\mathcal{L}_{T,(q,r_\perp^*)}$
6:     **end if**
7:     **if** $\mathcal{L}_{T,(q,r_\perp^*)}$ has not visited lists **then**
8:         Let $L$ be a not visited lists from $\mathcal{L}_{T,(q,r_\perp^*)}$
9:         Mark $L$ as visited
10:         **return** $L$
11:     **else**
12:         **return** $\emptyset$
13:     **end if**
14: **end if**

**Note 1:** At any moment, if it is possible (e.g. at line 5), $r_\top^*$ is updated to a tighter bound, in such case $q^*$ must be updated too.

---

Please notice that the efficiency of next_best($T$) is tied to the particular implementations. For example, when $T$ is a tree, next_best($T$) procedure can be implemented using a stack to emulate recursive calls.

*Example 1 (next_best($T$) over a single pivot).* Consider a pivot $P \in S$, inducing a partition $\Pi_P$, using a discretizing function $g(d(P, u))$ for each $u \in S$. In the figure 1 exemplifies this with $|\Pi_P| = 8$.

The first step is to find the list which may contain $q$ with radius zero, i.e. $|g(d(q, P)) - g(d(u, P))| \leq g(r_\perp^*) = 0$. Then, we must increase $r_\perp^*$ such that we advance in to the next promising list, e.g. $g(r_\perp^*) = 1$. In the process both $r_\top^*$ and $r_\perp^*$ are adjusted. The process is repeated until both bounds meet. Figure 1 depicts the advance of next_best($T$). Please notice that this procedure is the one used by any pivot based index.

*Example 2 (next_best($T$) for the LC).* We start the procedure by determining the order to review the lists. In the process, $r_\top^*$ bound is improved. This step evaluates $d(q, c_i)$ for all $c_i \in C$, sorting $c_i$'s in ascending order of $|d(q, c_i) - \text{cov}(c_i)|$, because this is the reviewing order induced as $r_\perp^*$ increases. Section 1.2 shows the details of searching in an LC.

In general, k-nn searching is implemented replacing $q^*$ as a priority queue of fixed cardinality $k$. In this variation, both $r_\perp^*$ and $r_\top^*$ bound the covering radius of the $k$ nearest neighbor.

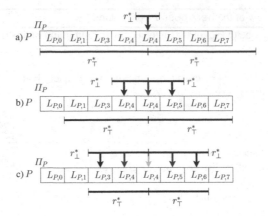

**Fig. 1.** Next best procedure for a single pivot $P$

## 4.1 Expected Performance

The cost of solving a query, the size of the candidate list, is closely related to the expected performance of the internal indexes. A rule of thumb is that every index must provide a diverse (low correlated) set of candidates, such that $\bigcap_{T \in \Lambda} \mathcal{C}_T$ is close to the query answer, i.e. items not being in the result set should be discarded by at least one index in $\Lambda$.

Let $P_{T,u}$ be the probability that a random object $u$ needs to be reviewed for some query $(q, r)_d$ in the metric index $T$. Similarly, let us define $P_{\Lambda,u}$ for the set of indexes in $\Lambda$ as the probability that $u$ cannot be discarded by an index $\Lambda$. Assume all $P_{T,u}$ for $T \in \Lambda$ are independent probabilities, then $P_{\Lambda,u} = \prod_{T \in \Lambda} P_{T,u}$. If each $T$ has been constructed ensuring that $P_{x,u} \simeq P_{y,u}$ for all $x, y \in \Lambda$ then $P_{\Lambda,u} = P_{x,u}^\lambda$. This suggest that we can improve our search simply adding (independent) indexes to $\Lambda$, arbitrarily decreasing $P_{\Lambda,u}$. This simplification can be seen as a probabilistic lower bound in the computing of the probability $P_{\Lambda,u}$.

Since $P_{\Lambda,u} = |\bigcap_{T \in \Lambda} \mathcal{C}_T|/n$, the probability is not lower bounded, because the intersection could be empty. On the other hand, the upper bound is $P_{\Lambda,u} = \min_{T \in \Lambda} |\mathcal{C}_T|/n$. The probabilistic lower bound is found for independent probabilities under some probability distribution. A more precise model may consider the dependency between objects and the characteristics of the indexes. As a formative example let $T_i$ and $T_j$ be LC indexes in the collection, then one of the following cases may arise:

– $u$ is a center on $T_i$ and $v \in I_u$, then if $v$ is a center on $T_j$ it is possible that $u \in I_v$.
– $u, v \in I_c$ for a center $c$ on $T_i$, then if $c'$ is a center on $T_j$, it is possible that $u, v \in I_{c'}$.
– the most common case is that a query ball intersects with centers and buckets such that the previous cases are extended to set of centers and buckets.

With the searching algorithms described, we need to speed up the LC construction.

## 4.2 Revisiting the LC

The LC is an efficient metric index, but it has an expensive preprocessing time complexity for high intrinsic dimensional datasets. One observation is that the original algorithm do not fix the order in which centers are selected. In other words, center selection do not make sense because the centers are selected sequentially starting from a seed center. In the original paper the authors propose four heuristics for iteratively selecting the centers.

One of the requirements of the PMI is to have the partitions selected independently, see section 4.1. For this reason we introduce a new randomized construction of the LC. This serve two goals. Firstly, the construction is faster, and on the other hand the diversity on the partitions (as required by PMI) is achieved.

As explained in section 4.1, our method requires a high diversity in the partitions of the underlying indexes, hence we must promote this behavior. In this sense, please notice that the original LC does not necessarily select $c$ randomly.

Our contribution replace the *deterministic* selection of the center by a *random* selection of $c \in S$. This modification is implemented applying Knuth's Fisher-Yates shuffle to the set of identifiers. The complexity remains the same.

The recipe for high dimensional datasets needs $n/m = O(1)$, that is prohibitive for many real world applications since the preprocessing time of LC, $O(nm)$, becomes $O(n^2)$. An alternative strategy is to produce non optimal LCs, such that its preprocessing step would be cheaper. As we will show experimentally, this non optimal construction does not affect our index since the combination of several suboptimal LC's produces a faster index than a single optimal LC.

Let $m = O(\log^b n)$ for some $b \geq 1$. Under this approach, we require close to $nm/2 = O(n \log^b n)$ distance computations. If $b = 1$ then we obtain $O(n \log n)$ time, similarly to VPT or BKT [3].

Another possible approach is to define $m = O(n^\beta)$, resulting in a preprocessing step of $O(n^{1+\beta})$, and $n/m = O(n^{1-\beta})$.

*Example 3 (Gaining three orders of magnitude).* Let $n = 10^6$, suppose that $m = O(\log^b n)$, specifically $b = 2$, and an involved constant of 2.52, thus $m \sim 1000$. Finally, $nm/2$ becomes $5 \times 10^8$, which is much smaller than $5 \times 10^{11}$.

The same preprocessing cost for this example configuration is found fixing $m = n^{1/2}$, such that $n/m \simeq O(\sqrt{n})$.

Using these configurations, the number of distances needed to search are, a priori, larger than the required by an optimal LC if using a single index. When using several indexes our PMI reduces the number of distance computations over the optimal LC. In the following sections we experimentally verify our claims, and obtain very good tradeoffs among space, search time, and preprocessing time.

## 5 Experimental Results

We conducted experiments over synthetic and real-world datasets. Synthetic data are randomly generated vectors in the unitary cube, these datasets are used to describe the characteristics of the LC varying the intrinsic dimension. Real-world datasets are used to show the performance as will be found in databases obtained from real-world processes.

### 5.1 Description of Datasets

- Random vectors (RVEC). Six randomly generated sets of vectors in the unitary cube with dimensions 4, 8, 12, 16, 20, and 24 and fixed size of $n = 10^6$. Two hundred nearest neighbor queries form our query sets. Each query object is a randomly generated vector of the dimension of the dataset.
- Colors. A real world benchmark of 112682 color histograms (vectors of 112 co-ordinates) with $L_2$ as distance. The source of this database is the *sisap project* (http://www.sisap.org). Each query $q$ is composed of two random objects $u, v$ in the dataset, such that $q_i = (u_i + v_i)/2$. Queries are also for nearest neighbor.
- CoPhIR-1M. This database consists of 1 million of objects selected from the CoPhIR project [29]. Each object is a 208-dimensional vector and we use the $L_1$ distance. Each vector was created as a linear combination of five different MPEG7 vectors as described in [29]. We selected 200 vectors (they were not indexed) as queries. Queries are also for nearest neighbors.

All the algorithms were written in C#, with the Mono framework (http://www.mono-project.org). Algorithms and indexes are available as open source software in the *natix* library (http://www.natix.org). The experimentation was carried in a four quad-core Intel Xeon 2.40 GHz workstation with 32 GiB of RAM, running CentOS Linux. The entire databases and indexes are maintained in main memory and without exploiting any parallel capabilities of the workstation.

Please notice that the build time is critical for the LC, since it requires $O(nm)$ distance computations, and high dimensional datasets require $n/m = O(1)$, we got a $O(n^2)$ distance computations. Under this perspective LC is limited to lightweight distances or small databases. Our approach is based on exploiting configurations $m = o(n)$, such as $m = O(n^\beta)$ with some $0 < \beta < 1$, and to take advantage of several indexes and the diversity found in their partitions.

We provide an extensive comparison against the original List of Clusters, however we avoid a larger comparison against other structures since LC is a typical baseline for metric indexes. Also, we avoid disk based indexes like M-Tree or PM-Tree since they are commonly worried about I/O efficiency, which is not directly comparable with our approach.

### 5.2 Build Time

In order to discover the gain in the preprocessing time, table 1 shows the real time to build a single instance of an index for RVEC-4, $n = 10^6$. In this setup, the distance function has a low-cost.

It is interesting to notice the preprocessing speedup achieved by simply increasing $n/m$. The preprocessing time for $n/m = 1024$ is 6.2 times faster than the LC with $n/m = 128$, and 48.8 times than $n/m = 16$. This relation from 128 to 16 is 7.9. This implies that creating several $\lambda$ indexes is even cheaper than create a single *optimal* LC.

**Table 1.** Preprocessing time for RVEC-4 and $n = 10^6$

| method | $n/m$ | $m$ | preprocessing time | |
|---|---|---|---|---|
| | | | seconds | human readable |
| LC | 1024 | 976 | 331.13 | 5 min 31.13 sec. |
| LC | 128 | 7812 | 2056.5 | 34 min 16.52 sec. |
| LC | 16 | 62500 | 16163.16 | 4 hours 29 min. |

Furthermore, as indexes are independently created, they can be built in parallel, such that the required time is close to the built time of a single instance plus the time to put them together. This later time is negligible.

## 5.3  Searching Performance

The complexity measured as number of computed distances is useful to extrapolate to other kind of distances; independently of the tested hardware, and of the database. On the other hand, the real time is necessary to measure the method to be used in practical applications. We are interested in both parameters.

Figure 2, depicts both the average computed distances and the required time for a single nearest neighbor query in the RVEC set of databases. The curves with $\lambda = 1$ are equivalent to the performance of the LC in the specified configuration. In these series of plots, there are three variables: $n/m$ (a pair of plots per row), $\lambda$ (a curve per value), and the dimension (the horizontal axis). The first column shows the cost presented as number of computed distances, figures 2(a), 2(c), and 2(e) respectively for $n/m$ of 16, 128, and 1024. We can observe that for large dimensionality (dimension 24), the PMI is unbeatable, particularly for $n/m = 128$, where we need to review 20% of the database, compared against 38% of the best configuration achieved with the LC ($n/m = 16$). For lower dimensions, it is natural to select large $n/m$, for example, the PMI requires $\lambda = 2$ and $n/m = 1024$ to review 0.3% of the database for dimension 4 (figure 2(e)) while the same $\lambda$ review 10% of the database for $n/m = 16$. A similar proportion is found for the plain LC ($\lambda = 1$). In general, we always can construct several indexes with large $n/m$ and increase $\lambda$ as required by the expected performance, obtaining that performance increase as $\lambda$ growth as observed in the left column of figure 2. The real search times (right column figure 2) are not showing the same dramatic variations, since they reflect the cost of the union-intersection algorithms and some effects of the cache. Nevertheless, the speed up introduced by the PMI is noticiable, as shown by figures 2(d) and 2(f).

Figure 3 shows performances for our real world databases, Colors and CoPhIR-1M. The experiment shows the dependency of the performance with $n/m$, i.e. the main parameter of the practical List of Clusters, and both average total (internal + external) number of distances and real time (figures 3(a) and 3(b)) required to solve a nearest neighbor query. As in previous figures, the left column show the number of distance computations, and the right one is showing the real time. Here the performance boost occurs varying $n/m$ and $\lambda$, for the fixed dimension.

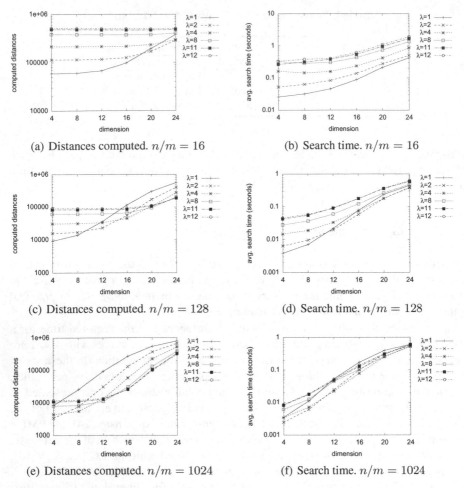

**Fig. 2.** Behavior of the PMI using LC searching for the nearest neighbor with a varying increasing intrinsic dimension over RVEC databases

The number of distance computations is optimized for LC on Colors at $n/m = 32$, and until this bucket size, the plain LC is the best option. For larger $n/m$ values, there exists a speed up (distance computations and real time) for all $\lambda$ values, with special remark on $n/m$ larger than 128 and $\lambda \geq 4$. On these setups the cost is half of the best LC. Furthermore, they have smaller preprocessing time. On CoPhIR-1M, the performance is quite similar, but we must remark that the preprocessing time imply an enormous difference since $n = 10^6$ and each vector contains 208 coordinates, which is very costly as preprocessing time. The LC ($\lambda = 1$) is optimized at $n/m = 128$ after this value, all setups are better than the single LC, in both the distance computations and the real time, see figures 3(c) and 3(d). We must remark that we found similar cost for $n/m = 64$, but we do not show such smaller values since preprocessing time is much larger and it does not improves neither LC nor PMI.

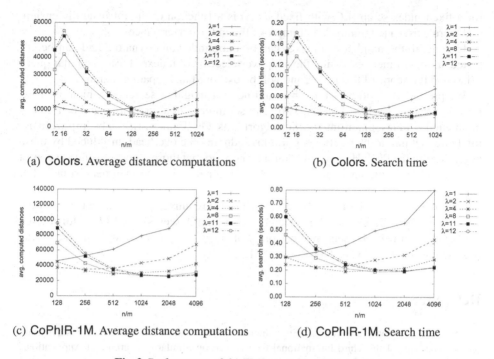

(a) **Colors.** Average distance computations                 (b) **Colors.** Search time

(c) **CoPhIR-1M.** Average distance computations         (d) **CoPhIR-1M.** Search time

**Fig. 3.** Performance of the PMI on real world datasets

## 6    Conclusions and Future Work

We presented a new metric index for general metric spaces called the Polyphasic Metric Index (PMI) that is more robust to the dimension that the well known List of Clusters (LC), one of the most robust indexes with small memory requirements. The central idea of our index is the usage of several backend indexes, where each one respond with a set of candidates containing the exact result set of the proximity query. The final set of candidates is obtained by intersecting all individual sets. We choose the List of Cluster index as backend index. This selection is driven by the fact that the LC is a fast, and small metric index. Those properties are inherited by our index, and even when our index is composed of several LC backend indexes, their configuration allows a very fast preprocessing time, far from the $O(n^2)$ time required by the original version of the algorithm. For example, we obtain faster searches than the LC with $O(n^{1.5})$ preprocessing time. Due to the compound functionality of our index, it is possible to adjust the number of indexes at searching time. Such that *hard* queries are solved with a complex machinery (several indexes), and *easy* ones with a simpler setup (few or one index).

The above scheme is easily adapted to discover the better PMI configuration for the (unknown) intrinsic dimension of the datasets. Based in our experimental evidence, configurations with large $n/m$ are quite good for small dimensions, and several backed indexes with this setup are useful for high intrinsic dimensions. We conjecture that $\lambda$

for a fixed index setup (LC with fixed $n/m$) is a function of the intrinsic dimension, and there exists a maximum $\lambda$ that improve the index performance.

In general, the number of indexes can be unbounded and dynamic and can be adjusted at query time. Discussing the adaptive selection of indexes for a particular query is beyond the scope of this paper, and will be explored in a separate manuscript.

We presented algorithms for range and nearest neighbor searches, which are new. Both algorithms are based on set union and set intersection operations and can be implemented with fast union-intersection algorithms. Our framework can be improved using better set union-intersection algorithms reducing the overhead introduced by these operations in the PMI. However it should be noticed that those algorithms must support *partial intersections*, since they are the core of our iterative, optimal nearest neighbor algorithm.

The major drawback of the PMI approach is the required space, which is a multiple of $\lambda$, even when the LC is a light weight index and $\lambda$ seems to be $O(1)$ for a fixed dataset. This drawback opens the possibility of applying compression techniques to the representation of the LC and the entire PMI.

# References

1. Skopal, T.: Where are you heading, metric access methods?: a provocative survey. In: Proceedings of the Third International Conference on Similarity Search and Applications, SISAP 2010, pp. 13–21. ACM, New York (2010)
2. Samet, H.: Foundations of Multidimensional and Metric Data Structures, 1st edn. The morgan Kaufman Series in Computer Graphics and Geometic Modeling. Morgan Kaufmann Publishers, University of Maryland at College Park (2006)
3. Chávez, E., Navarro, G., Baeza-Yates, R., Marroquín, J.L.: Searching in metric spaces. ACM Comput. Surv. 33(3), 273–321 (2001)
4. Böhm, C., Berchtold, S., Keim, D.A.: Searching in high-dimensional spaces: Index structures for improving the performance of multimedia databases. ACM Computing Surveys 33(3), 322–373 (2001)
5. Zezula, P., Amato, G., Dohnal, V., Batko, M.: Similarity Search: The Metric Space Approach (Advances in Database Systems). Springer-Verlag New York, Inc., Secaucus (2005)
6. Pestov, V.: Intrinsic dimension of a dataset: what properties does one expect? In: Proc. 20th Int. Joint Conf. on Neural Networks, Orlando, FL, pp. 1775–1780 (2007)
7. Pestov, V.: An axiomatic approach to intrinsic dimension of a dataset. Neural Networks 21(2-3), 204–213 (2008)
8. Pestov, V.: Indexability, concentration, and vc theory. In: Proceedings of the Third International Conference on Similarity Search and Applications, SISAP 2010, pp. 3–12. ACM, New York (2010)
9. Patella, M., Ciaccia, P.: Approximate similarity search: A multi-faceted problem. Journal of Discrete Algorithms 7(1), 36–48 (2009)
10. Zezula, P., Amato, G., Dohnal, V., Batko, M.: Similarity Search - The Metric Space Approach, 1st edn. Advances in Database Systems, vol. 32. Springer (2006)
11. Amato, G., Rabitti, F., Savino, P., Zezula, P.: Region proximity in metric spaces and its use for approximate similarity search. ACM Trans. Inf. Syst. 21, 192–227 (2003)
12. Tellez, E.S., Chávez, E., Navarro, G.: Succinct nearest neighbor search. In: Proc. 4th International Workshop on Similarity Search and Applications (SISAP). ACM Press (2011)

13. Tellez, E.S., Chavez, E., Graff, M.: Scalable Pattern Search Analysis. In: Martínez-Trinidad, J.F., Carrasco-Ochoa, J.A., Ben-Youssef Brants, C., Hancock, E.R. (eds.) MCPR 2011. LNCS, vol. 6718, pp. 75–84. Springer, Heidelberg (2011)

14. Chavez, E., Figueroa, K., Navarro, G.: Effective proximity retrieval by ordering permutations. IEEE Transactions on Pattern Analysis and Machine Intelligence 30(9), 1647–1658 (2008)

15. Amato, G., Savino, P.: Approximate similarity search in metric spaces using inverted files. In: InfoScale 2008: Proceedings of the 3rd International Conference on Scalable Information Systems, ICST, Brussels, Belgium, pp. 1–10. ICST (Institute for Computer Sciences, Social-Informatics and Telecommunications Engineering) (2008)

16. Esuli, A.: Pp-index: Using permutation prefixes for efficient and scalable approximate similarity search. In: Proceedings of the 7th Workshop on Large-Scale Distributed Systems for Information Retrieval (LSDS-IR 2009), Boston, USA, pp. 17–24 (2009)

17. Gionis, A., Indyk, P., Motwani, R.: Similarity search in high dimensions via hashing. In: VLDB 1999: Proceedings of the 25th International Conference on Very Large Data Bases, pp. 518–529. Morgan Kaufmann Publishers Inc., San Francisco (1999)

18. Andoni, A., Indyk, P.: Near-optimal hashing algorithms for approximate nearest neighbor in high dimensions. Communications ACM 51(1), 117–122 (2008)

19. Chávez, E., Navarro, G.: A compact space decomposition for effective metric indexing. Pattern Recogn. Lett. 26, 1363–1376 (2005)

20. Kyselak, M., Novak, D., Zezula, P.: Stabilizing the recall in similarity search. In: Proceedings of the Fourth International Conference on Similarity Search and Applications, SISAP 2011, pp. 43–49. ACM, New York (2011)

21. Yianilos, P.N.: Excluded middle vantage point forests for nearest neighbor search. Technical report, NEC Research Institute, Princeton, NJ (July 1998)

22. Skopal, T.: Pivoting m-tree: A metric access method for efficient similarity search. In: DATESO 2004, pp. 27–37 (2004)

23. Ciaccia, P., Patella, M., Zezula, P.: M-tree: An efficient access method for similarity search in metric spaces. In: Proceedings of the 23rd International Conference on Very Large Data Bases, VLDB 1997, pp. 426–435. Morgan Kaufmann Publishers Inc., San Francisco (1997)

24. Micó, M.L., Oncina, J., Vidal, E.: A new version of the nearest-neighbour approximating and eliminating search algorithm (aesa) with linear preprocessing time and memory requirements. Pattern Recogn. Lett. 15, 9–17 (1994)

25. Hwang, F.K., Lin, S.: A simple algorithm for merging two disjoint linearly ordered sets. SIAM Journal Computing 1(1), 31–40 (1972)

26. Baeza-Yates, R.A.: A fast set intersection algorithm for sorted sequences. In: CPM, pp. 400–408 (2004)

27. Demaine, E.D., López-Ortiz, A., Munro, J.I.: Adaptive set intersections, unions, and differences. In: SODA 2000: Proceedings of the Eleventh Annual ACM-SIAM Symposium on Discrete Algorithms, pp. 743–752. Society for Industrial and Applied Mathematics, Philadelphia (2000)

28. Barbay, J., Kenyon, C.: Adaptive intersection and t-threshold problems. In: Proceedings of the 13th ACM-SIAM Symposium on Discrete Algorithms (SODA), pp. 390–399. ACM-SIAM, ACM (January 2002)

29. Bolettieri, P., Esuli, A., Falchi, F., Lucchese, C., Perego, R., Piccioli, T., Rabitti, F.: CoPhIR: a test collection for content-based image retrieval. CoRR abs/0905.4627v2 (2009)

# Efficient Similarity Search in Metric Spaces with Cluster Reduction*

Luis G. Ares, Nieves R. Brisaboa, Alberto Ordóñez Pereira, and Oscar Pedreira

Database Laboratory, Universidade da Coruña
Campus de Elviña s/n, 15071, A Coruña, Spain
{lgares,brisaboa,alberto.ordonez,opedreira}@udc.es

**Abstract.** Clustering-based methods for searching in metric spaces partition the space into a set of disjoint clusters. When solving a query, some clusters are discarded without comparing them with the query object, and clusters that can not be discarded are searched exhaustively. In this paper we propose a new strategy and algorithms for clustering-based methods that avoid the exhaustive search within clusters that can not be discarded, at the cost of some extra information in the index. This new strategy is based on progressively reducing the cluster until it can be discarded from the result. We refer to this approach as *cluster reduction*. We present the algorithms for range and *k*NN search. The results obtained in an experimental evaluation with synthetic and real collections show that the search cost can be reduced by a 13% - 25% approximately with respect to existing methods.

**Keywords:** similarity search, metric spaces, cluster reduction.

## 1 Introduction

Similarity search is a typical operation in many areas of computer science, such as pattern recognition, computational biology, multimedia information retrieval, or recommender systems, to name a few. Given a collection of objects and a function measuring the distance or dissimilarity between any two of them, similarity search finds the most similar objects to another one given as a query. The comparison of two objects is supposed to be computationally costly, so the goal of metric access methods (MAMs) is to solve the queries with the minimum number of distance evaluations.

Similarity search can be formalized through the mathematical concept of metric space. A metric space is composed by an universe of objects and a metric that determines the distance or dissimilarity between any two objects of that universe. Methods for searching in metric spaces preprocess the collection and

---

* This work has been partially funded by "Ministerio de Ciencia y Innovación" (PGE and FEDER) refs. TIN2009-14560-C03-02, TIN2010-21246-C02-01, and ref. AP2010-6038 (FPU Program) for Alberto Ordóñez Pereira, and by "Xunta de Galicia" refs. 2010/17 (Fondos FEDER), and 10SIN028E.

build indexes that store precomputed information about the objects in collection. This information is used during the search together with the properties of metric spaces to prune the search space and thus compare the query with a small portion of the objects in the collection [1–3].

The most studied types of query in metric spaces are range search and $k$NN search. *Range search*, $R(q, r)$, obtains all the objects up to a distance $r$ from the query object $q$. *Near neighbor search*, $kNN(q)$, retrieves the $k$ most similar objects to the query.

Methods for searching in metric spaces can be classified in pivot-based methods and clustering-based methods [1]. *Pivot-based methods* select a subset of objects from the collection to be used as pivots, and the index stores the distances from the pivots to the rest of objects. Given a query, the query object is compared with the pivots, and these distances are used to discard as many objects as possible without comparing them with the query.

*Clustering-based methods* partition the space into a set of disjoint clusters. Each cluster is represented by an object used as the cluster center. Each object in the collection belongs to the cluster corresponding to its closest center. The index maintains the information of each cluster, typically its center and its *covering radius*, that is, the distance from the cluster center to its furthest object in the cluster. Given a query, the query object is compared with the center of each cluster, and complete clusters are discarded if the information provided by the index determines that they can not contain objects in the result set. If a cluster can not be discarded, it is searched exhaustively, that is, the query object is compared with all the objects that belong to that cluster.

In this paper we present a new strategy and algorithms for precise metric-based search that avoid the exhaustive search within a cluster that can not be discarded by the index. Our proposal is based on the idea of defining regions within each cluster with respect to its center, in such a way that when we search within the cluster, it can be progressively reduced by discarding some of its regions, until the rest of the cluster is completely discarded. We refer to this strategy as *cluster reduction*, and it can be applied in any method using the covering-radius pruning criteria for discarding objects. Although we need to store more information in the index to maintain the regions within each cluster, we show that the space complexity of the index remains $O(n)$. We present the algorithms for range search and $k$NN search, and an experimental evaluation with real and synthetic collections of different nature that shows that this approach can improve the search performance by a 13% - 25% approximately with respect to existing methods. The results presented in the experimental evaluation also consider the effect of cluster reduction on the size of the index, and the trade-off between search cost improvement and space requirements.

The rest of the paper is structured as follows. Next Section briefly reviews related work on clustering-based methods for metric spaces. Section 3 presents cluster reduction and the algorithms for range and $k$NN search. In Section 4 we present the results we obtained from the experimental evaluation. Finally, Section 5 summarizes the conclusions of the paper and lines for future work.

## 2   Related Work

A *metric space* is a pair $(X, d)$, where $X$ is an *universe* of objects, and the function $d : X \times X \longrightarrow \mathbb{R}^+$ is a *metric* measuring the dissimilarity $d(x, y)$ between any two objects $x, y \in X$. The metric holds the properties of positiveness $(d(x, y) \geq 0)$, symmetry $(d(x, y) = d(y, x))$, and the triangle inequality $(d(x, y) \leq d(x, z) + d(z, y))$. The database or *collection* of objects is a finite subset $S \subseteq X$ of size $|S| = n$.

Clustering-based methods partition the space into a set of disjoint clusters, and the index stores the information about this partition. The information for cluster $\mathcal{C}_i$ includes at least the object used as the center of the cluster, $c_i$, the set of objects that belong to that cluster, and the covering radius, $cr_i = max\{d(c_i, x)/x \in \mathcal{C}_i\}$, that is, the distance from the center to its furthest object in the cluster. For each cluster $\mathcal{C}_i$, its center and its covering radius define a *ball* in the space, $(c_i, cr_i)$, containing all the objects that belong to the cluster.

Given a range query $R(q, r)$, the query object is compared with each cluster center $c_i$, obtaining the distances $d(q, c_i)$. A range query defines a ball in the space, $(q, r)$. For each cluster $\mathcal{C}_i$, the cluster can be discarded without comparing the query with its objects if $d(q, c_i) - cr_i > r$, that is, if the ball $(c_i, cr_i)$ does not intersect the ball $(q, r)$.

In the case of $k$NN queries, the distance $d(q, c_i) - cr_i$ gives us a lower bound on the distance from $q$ to any of the objects that belong to the cluster $\mathcal{C}_i$, that is, $\forall x \in \mathcal{C}_i$, $d(q, x) \geq d(q, c_i) - cr_i$ (if the query object $q$ falls within the ball defined by the cluster, $(c_i, cr_i)$, the lower bound is negative, so it does not give us useful information). This lower bound gives us a hint of which clusters we should visit first and when to stop the search, that is, when none of the remaining clusters contains objects closer to $q$ than its current $k^{th}$ nearest neighbor.

The search complexity is given by the sum of the internal and external complexities. The *internal complexity* is the number of distance evaluations needed to compare the query object with the cluster centers. The *external complexity* is the number of distance evaluations needed to compare the query object with the objects in the clusters that could not be discarded.

Clustering-based methods build smaller indexes and behave better in high-dimensional spaces with respect to pivot-based methods. Existing clustering-based methods differ in how they partition the space, how that partitioning is reflected in the index structure, and on the criteria used for pruning the search space. Most methods partition the space in a recursive way and create a tree index that reflects that partition.

BST [4] recursively partitions each cluster into two clusters, and creates a tree index that maintains the information of the partition. In a first level, two objects are selected as cluster centers. The root of the index stores the center and covering radius of each cluster, and each cluster is then recursively partitioned following the same schema. GHT [5] partitions the space following the same schema, but changes the criteria to prune the tree during the search. In this case, the left subtree at each node is searched if $d(q, c_l) - r \leq d(q, c_r) + r$, and the right subtree is searched if $d(q, c_r) - r \leq d(q, c_l) + r$. GNAT [6] is a generalization of GHT in which more than two clusters are created at each node. In addition, each node stores the distances between the centers of the clusters, so some of them can be discarded without comparing them with the query object. VT [7] improves BST by using two or three centers in each node and storing in each new node the closest object from the parent node. SAT [8] creates a tree structure that approximates the *Delaunay graph* of the partition, and the search traverses the tree discarding complete clusters when possible.

The M-Tree [9] recursively partitions the space trying to obtain clusters as compact as possible to improve the search cost. This method established an important landmark since it supports the dynamics of a real database system. Its structure is suitable for efficient secondary memory storage, and it supports dynamic insertions and deletions of objects in the database without degrading the index performance, by rearranging the index on such operations, and adapting it to the new content of the database. The Slim-Tree [10] is a well-known modification of the M-Tree that reduces the overlap between the clusters.

All the methods described above create a recursive partition that generates a tree-like index. List of Clusters [11] follows a different approach and organizes the index as a list instead of as a tree. A first object is used as a cluster center for the first cluster. Once this cluster is full, the rest of the collection is processed in the same way. The size of the cluster is determined either by a fixed covering radius or by a fixed number of objects. The search traverses the list discarding the clusters when possible, and exhaustively searching non-discarded clusters.

## 3   Similarity Search with Cluster Reduction

In this Section we present the idea of cluster reduction, and the corresponding algorithms for range search and $k$NN search. The goal of cluster reduction is to decrease the external complexity of the search by avoiding the exhaustive search within clusters that could not be discarded with the information of the index.

We present the algorithms for the particular case of List of Clusters, although this approach could be extended to other clustering-based methods using the covering radius pruning criteria.

### 3.1   Defining Intermediate Regions

Let $\mathcal{C}_i$ be a cluster with center $c_i$ and covering radius $cr_i$, and let $\{x_{i1}, \ldots, x_{im}\}$ be the set of objects that belong to $\mathcal{C}_i$ (where $m$ is the size of the cluster). These objects define a set of distances with respect to the center of the cluster, $\{d(c_i, x_{i1}), \ldots, d(c_i, x_{im})\}$. We select some of these distances to define intermediate regions within the cluster $\mathcal{C}_i$, in such a way that each selected distance acts like an additional internal covering radius. If we want to define $\beta$ regions

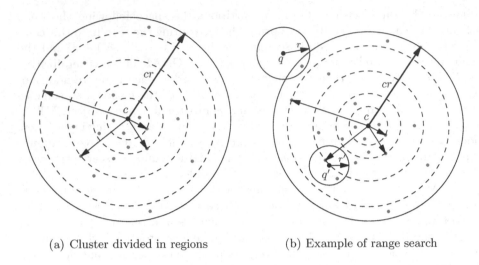

(a) Cluster divided in regions          (b) Example of range search

**Fig. 1.** Range search with cluster reduction

within each cluster, we have to select $\beta - 1$ distances. We select the distances in such a way that each region contains the same number of objects. If the number of regions, $\beta$, does not divide the number of objects in the cluster, one of the regions will have fewer objects than the others.

Figure 1(a) shows an example in which a cluster is divided into five internal regions. Each internal region contains four objects. The reason for selecting regions that contain the same number of objects is that in this way they adapt better to the distribution of the objects within the cluster. As previously stated in [1, 11], the distribution of the objects with respect to the center of the cluster is not uniform. Choosing regions with the same "width" could led to a region that contains almost all the objects in the cluster, which would be useless for the algorithms we present in this section.

The smallest number of regions we can define within a cluster is two, and, at most, we could define as many regions as objects in the cluster. In the first case, we have the chance of discarding half of the cluster without comparing it with the query. In the second case, each comparison of $q$ with an object of the cluster gives us a chance of discarding the rest of the cluster without comparing it with the query. Therefore, the more the regions we define within clusters, the more chances of reducing the search cost.

However, the number of regions also affects the space requirements of the index. If we define as many regions as objects in the cluster, the space requirements would be almost the double, although the space complexity would still be $O(n)$. Therefore, there is a trade-off between the search cost and space requirements in terms of the number of regions. As we will see in the experimental evaluation, a small number of regions leads to a search cost that is very close to that obtained when the number of regions equals the number of objects.

## 3.2   Additional Considerations

Cluster reduction can be viewed as a generalization of the range-pivot distance constraint policy for discarding objects [2] if we consider the limit of each internal region as an additional covering radius. Similarly, cluster reduction can be viewed as a generalization of the object-pivot distance constraint for discarding objects [2], since the center of the cluster assumes a role similar to that of a pivot for the objects of that cluster. By knowing to which region each object in the cluster belongs, and the limits of that region, we have a range bounding the distance from the cluster center to each object, which is a case of range coarsening [1].

The idea of defining regions within clusters has already appeared in previous work. All methods that recursively partition the space divide each cluster into smaller clusters, although the leaf nodes have to be searched exhaustively. VPT [5] and MVPT [12] recursively partition each cluster into regions of the same size, which are further partitioned using the same schema. M-Index [13] assigns to each object a *key* that combines the identifier of the cluster to which it belongs, and its distance to the center of the cluster. This can be seen as a particular case of cluster reduction in which the distances from all the objects to the cluster center are stored. The PM-Tree [14] is an extension of the M-Tree in which each cluster is divided into *hyper-rings* with respect to a set of pivots.

The main differences of our proposal with previous approaches are: ($i$) we divide the cluster into several concentric regions with respect to the center (that is, without using other cluster centers or pivots) in an *onion-like* style, that is, each region is contained by another region; ($ii$) as we will see in the description of the algorithms for range search and $k$NN search, this allows us to process each cluster one region at a time until one of them allows us to discard the rest of the cluster; ($iii$) the results of our experimental evaluation show that using a small number of regions within each cluster leads to a search cost very close to that obtained when using as many regions as objects in the cluster.

Applying the cluster reduction strategy to an existing method modifies the structure of the index, since we need to store, for each cluster, which objects belong to each region, and also the distances from the center to the objects that define the limit of each region. These features of the index allow us to implement it using different data structures. Dynamic capabilities are also supported, allowing deletions (in general, logical deletions), and insertion of new objects. If the number of modifications in a cluster is high, restructuring the regions of the cluster in the index is necessary to maintain its search performance.

## 3.3   Range Search

The algorithm for range search proceeds initially in the same way as the original algorithm, only changing how clusters that can not be discarded are searched. Given a range query $R(q, r)$, the query object is compared with all the cluster centers $c_i$, obtaining the distances $d(q, c_i)$. For each cluster, we have three possibilities:

(a) $d(q, c_i) - cr_i > r$: in this case, the ball $(c_i, cr_i)$ defined by the cluster $C_i$ does not intersect the ball $(q, r)$ defined by the query, so the whole cluster $C_i$ is discarded from the result without comparing it with the query object.

(b) $d(q, c_i) - cr_i \leq r$, and not $d(q, c_i) + r \leq cr_i$: the ball $(c_i, cr_i)$ defined by $C_i$ intersects the ball $(q, r)$ defined by the query. In this case, we can not discard the cluster. Instead of comparing the query object with all the objects in the cluster, we start the comparison at the outermost region of the cluster. In each region, we compare the query object with all the objects in that region. As we process each region, the cluster is being progressively reduced. The search within the cluster stops when the limit of the next region does not intersect the query ball. That is, we use the limit of each region as the covering radius of the cluster as we reduce it.

(c) $d(q, c_i) - cr_i \leq r$, and $d(q, c_i) + r \leq cr_i$: the ball defined by the cluster not only intersects the query ball, but it contains it. In this case, the rest of the clusters do not have to be explored. The search within the cluster starts at the outermost region that intersects the query ball, and continues until we reach a region that does not intersect it.

Figure 1 shows an example of range search with cluster reduction. The left part of the figure shows a cluster containing 20 objects, that has been divided into 5 regions, each of them containing 4 objects. The arrows that start at the center of the cluster point to the objects that define the limit of each region. The right part of the figure shows two range queries.

In the first case, $R(q, r)$, the query ball intersects the cluster but is not contained within it. Therefore, the search compares the query with the objects in the outermost region of the cluster. Once we have processed this region and therefore reduced the cluster, the next region does not intersect the query ball, so the rest of the cluster can be discarded. In this example, the cost of searching within the cluster is reduced to a 20% of its objects.

In the second case, $R(q', r')$, the ball defined by the cluster contains the query ball. Since the query ball only intersects the second and third regions of the cluster (counting from the outermost one), the search can be reduced to those regions. The search within the cluster is solved comparing only the query object with two of the five regions. In this case, the search stops because we do not need to explore any other cluster.

### 3.4  $k$NN Search

In the case of $k$NN search, the algorithm proceeds by visiting the most promising clusters first, as in the original algorithm. The difference is that in each step of the search we do not process the whole cluster, but only its closest region to $q$ that has not still been processed. After processing that region, this cluster is reduced and the search continues with the next most promising cluster, including the cluster we have just processed and modified.

```
 1 kNNSearch(q)

 2 foreach cluster Cᵢ do
 3 |    dᵢ ← d(q, cᵢ)
 4 |    Insert(clustersQueue, Cᵢ, dᵢ − crᵢ)
 5 end

 6 Cᵢ ← pull(clustersQueue)
 7 while d(q, cᵢ) − crᵢ < radius(neighborsQueue) do
 8 |    Reduce(q, Cᵢ, neighborsQueue)
 9 |    Insert(clustersQueue, Cᵢ, dᵢ − crᵢ)
10 |    Cᵢ ← pull(clustersQueue)
11 end
```

**Algorithm 1.** Pseudocode for $k$NN search with cluster reduction

For each cluster $C_i$, the distance $d(q, c_i) - cr_i$ gives us a lower bound on the distance from $q$ to any object in $C_i$, that is:

$$\forall x \in C_i, \ d(q, x) \geq d(q, c_i) - cr_i \tag{1}$$

Given two clusters $C_i$ and $C_j$, the cluster $C_i$ is more promising than $C_j$ if:

$$d(q, c_i) - cr_i < d(q, c_j) - cr_j \tag{2}$$

Given a $k$NN query, the query object is compared with all the cluster centers, and the clusters are arranged into a priority queue, in such a way that the most promising clusters are processed first. In each step of the search, the most promising cluster is pulled from the queue, and the query is compared only with the objects in its closest region still not processed, updating the list of $k$ candidate nearest neighbors as necessary. The cluster is reduced since we have processed one of its regions, and it is reinserted again in the queue with a new priority, resulting from the reduction of the cluster. This procedure is repeated until the lower bound of the distance from $q$ to the next cluster to be processed is greater than the distance from $q$ to its current $k^{th}$ candidate neighbor.

Pseudocode 1 summarizes the algorithm for near neighbor search with cluster reduction. The algorithm uses two priority queues: *clustersQueue* stores the clusters according to how promising they are for the search, and *neighborsQueue* maintains in each step of the algorithm the $k$ nearest neighbors of $q$ among the objects that have already been processed (*neighborsQueue* has a limited capacity of $k$ objects). The function *Reduce* (line 8) compares $q$ with the objects in its closest region still not processed of $C_i$, and sets the covering radius of $C_i$ to the limit of its next region.

Figure 2 shows an example of 2NN search with two clusters. As we can see in the figure, the distance from $q$ to $C_1$, $d_1$, is smaller than the distance from $q$ to $C_2$, $d_2$, so $C_1$ is more promising and it is processed first. The query is compared with the objects in the first region of $C_1$, and the cluster is reduced and inserted again into the priority queue. Now, the distance from $q$ to $C_1$, $d_3$, is larger than

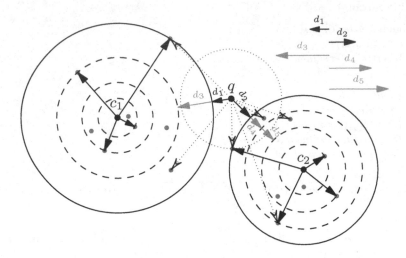

**Fig. 2.** $k$NN search with cluster reduction

the distance from $q$ to $\mathcal{C}_2$, $d_2$. In this point, the search stops processing $\mathcal{C}_1$ and jumps to $\mathcal{C}_2$, since it is more promising at this moment. After processing the first region of $\mathcal{C}_2$, the distance from $q$ to $\mathcal{C}_2$, now $d_4$, is smaller than the distance to $\mathcal{C}_1$, $d_3$, so we process the next region of $\mathcal{C}_2$. After processing the second region of $\mathcal{C}_2$, the search stops, since $d_3$ and $d_5$ are greater than the distance from $q$ to its second nearest neighbor (distance represented by the ball in dotted line). Note that the search has finished without having searched exhaustively within either $\mathcal{C}_1$ or $\mathcal{C}_2$, but only within one region of $\mathcal{C}_1$, and two regions of $\mathcal{C}_2$.

## 4    Experimental Evaluation

In this Section we present the results we obtained in the experimental evaluation of the new algorithms for range search and $k$NN search, considering both the effect of cluster reduction on the search cost and on the space requirements of the index.

We have implemented the new algorithms on a List of Clusters, under the framework provided by the SISAP Library [1]. In our experiments we used six datasets from the library:

- ENGLISH: a collection of $69,069$ words from the English dictionary.
- GERMAN: a collection of $75,086$ words from the German dictionary.
- NASA: contains $40,150$ images from NASA archives represented by feature vectors of dimension 20.

---

[1] http://sisap.org/Metric_Space_Library.html

- COLORS: contains 112, 544 color histograms, represented by feature vectors of dimension 112.
- UNIFORM-10, UNIFORM-12: collections of 100, 000 vectors of dimensions 10 and 12 respectively, with uniform distribution in the unitary cube.

We have chosen two collections of each type (words, images, and uniformly distributed vectors) since they contain objects of the same nature but present different complexities for the search. For each dataset, 90% of the objects were used as the collection to be indexed, and the remaining 10% were used as query objects. In the case of word datasets, objects were compared using the edit distance. In the case of vector datasets, we used the Euclidean distance.

## 4.1 Range Search

In order to evaluate the search cost obtained with cluster reduction for range search, we used cluster sizes $\{10, 20, 40, 60, 80, 100\}$, and values of $\beta$ (number of regions) of 2, 5, 10, and 20 regions within each cluster, as well as the case in which we used all possible regions within each cluster. The search radius was adjusted to retrieve an average of 0.01% of the database for each query. Figure 3 shows the mean number of distance computations for processing all the queries (10% of the dataset) for each collection and method. In the figure, "LC" stands for List of Clusters without applying cluster reduction, "LC CR-i" stands for list of clusters applying cluster reduction with $\beta = i$ regions within each cluster. "LC CR-all" stands for List of Clusters using as many internal regions as objects in each cluster.

As we can see in the results, the application of cluster reduction produces a significant improvement on the search cost. The higher cost improvement is obtained when we use all the objects in the cluster to define a region. However, the results obtained when using a smaller number of regions are very close to the best result. An important result is that a significant part of the search cost improvement is obtained when using just 5 regions within each cluster. Adding more regions within each cluster reduces the search cost even more, but from the results we can see that when using 10 or 20 regions within each cluster, adding more regions does not improve significantly. Note also that this behavior holds for all the cluster sizes we have considered and for all collections, no matter their size or complexity.

## 4.2 kNN Search

In order to evaluate the performance obtained with cluster reduction for kNN search, we used values of $k$ ranging from 1 to 10, and values of $\beta$ of 2, 5, 10 and 20 regions. We also considered the case in which each object in the cluster defines an internal region. In these experiments, the cluster size was fixed to 40, which produced good results in the previous set of experiments for all collections. Figure 4 shows the results we obtained. The legend of the figures follows the nomenclature we used in the experiments for range search.

As we can see in the results, the behavior of the search cost for different values of $\beta$ is similar to that obtained in the case of range search. The most significant improvement in the search cost is obtained when using just 5 regions within each cluster. When using 10 regions within each cluster, the search cost is very close to that obtained when maintaining the distances from the cluster center to all the objects in the cluster. The results are homogeneous for all values of $k$ in all collections.

## 4.3   Search Cost and Space Requirements

As we explained in Section 3, the number of regions defined within each cluster produces a trade-off between the improvement in search cost and the increment in the space requirements of the index. We obtain the best result in terms of search cost when all the objects in the cluster are used to define a region within the cluster, but this almost doubles the space requirements of the index. However, as we have already seen, when using a smaller number of regions within each cluster, the search cost is very close to the best result, with a very reduced increment in the space requirements.

Table 1 shows for each collection the size of the index obtained when using 2, 5, 10, 20, and all possible regions within each cluster (in Kbytes). The column "Relative" shows the relative value of the size of the index with respect to the list of clusters without cluster reduction (shown in first row, LC). The increments in the size of the index range from a 2% to a 20% when the number of regions is between 2 and 10. When each object of the cluster is used to define a region in the cluster, the space overhead reaches a 89%. Although in the case of using all possible regions the space of the index is almost doubled, cluster reduction preserves the $O(n)$ space complexity of the index, an important property of clustering-based methods when compared with pivot-based methods.

Figure 5 shows results on the trade-off between the search cost improvement and the increment of the index size for range search queries. The improvement of search cost and the increment of the size of the index are expressed as relative values (in %) with respect to the search cost and index size of the original list of clusters (as in Table 1). The figure considers all collections and values of $\beta$ in $\{2, 5, 10, 20, all\}$. When we define 2 regions within each cluster, the search cost is reduced a 10% approximately, with a space overhead of only a 2%. Using 5 regions produces a space overhead of 9%, while the search cost improvement ranges between 13% - 24% approximately with respect to list of clusters without cluster reduction, depending on the collection. When the number of regions is 10, the search cost improves slightly. From this point, using more regions within each cluster increments the size of the index but does not affect significantly to the search cost. This result is important since it shows that we only need to define a small number of regions within each cluster, independently of the size of the cluster.

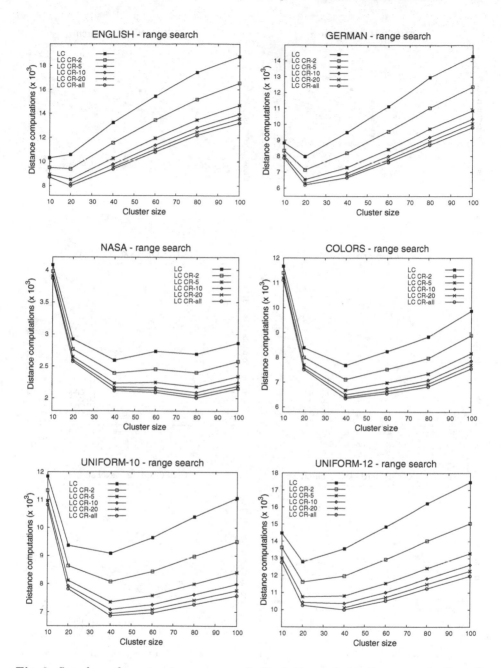

**Fig. 3.** Search performance in range search. Each figure shows the mean number of distance computations (in thousands of distances) in terms of the size of the clusters.

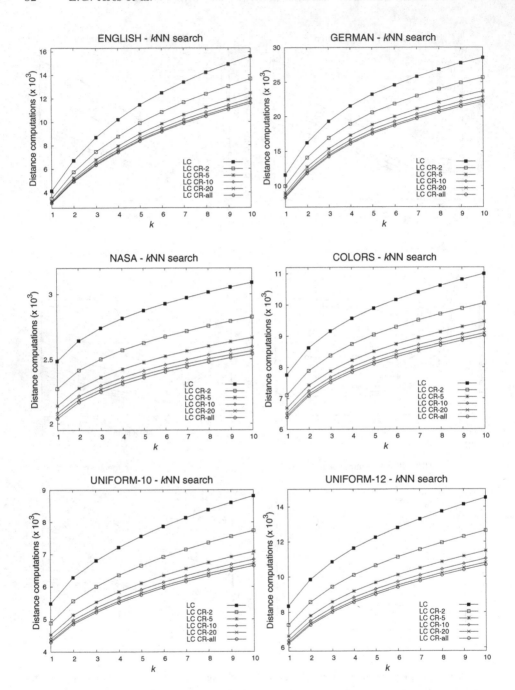

**Fig. 4.** Search performance in $k$NN search. Each figure shows the mean number of distance computations (in thousands of distances) in terms of $k$, the number of neighbors searched for each query object.

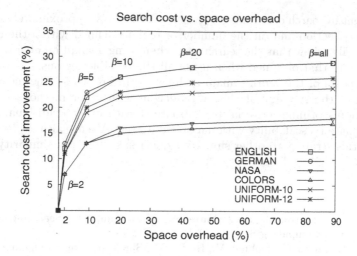

**Fig. 5.** Search cost vs. space trade-off in range search with cluster size 40

**Table 1.** Effect on space requirements (Kbytes)

| β | English | German | Nasa | Colors | Relative |
|------|---------|--------|--------|--------|----------|
| LC | 260.65 | 283.34 | 151.53 | 424.66 | 1.00 |
| 2 | 266.57 | 289.79 | 154.97 | 434.31 | 1.02 |
| 5 | 284.34 | 309.10 | 165.29 | 463.26 | 1.09 |
| 10 | 313.95 | 341.29 | 182.50 | 511.50 | 1.20 |
| 20 | 373.17 | 405.66 | 216.92 | 607.99 | 1.43 |
| all | 491.60 | 534.41 | 285.74 | 800.96 | 1.89 |

## 5 Conclusions

In this paper we have presented a new strategy, namely cluster reduction, and new algorithms that avoid the exhaustive comparison of the query with the objects in clusters that can not be discarded with the information in the index, thus reducing the external complexity. We have presented the algorithms for the particular case of List of Clusters, but this approach could be extended to other clustering-based methods using the covering radius pruning criteria.

We define internal regions within each cluster, in such a way that all regions contain approximately the same number of objects. When searching within a non-discarded cluster, the cluster is processed one region at a time, in order of proximity to the query. This allows us to progressively reduce the cluster until we can discard the rest of its regions. The definition of a reduced number of regions within each cluster reduces significantly the search cost, with a small increment in the size of the index, and maintaining the space complexity as $O(n)$.

We have presented an experimental evaluation with both real and synthetic collections from the SISAP Metric Spaces Library. Our results show that the improvement in search cost ranges between a 13% - 25% approximately depending on the collection and on the number of regions defined within the clusters. The results also show that the search cost when using a small number of regions is very close to that obtained when using all the possible regions.

Some aspects of this work remain as lines for future work. The most immediate, to test the strategy of cluster reduction with other methods using the covering radius pruning criteria, to compare it with other traditional MAMs, and to evaluate its scalability with massive data sets. We are also exploring how to extend this strategy to other similarity queries, such as the similarity join.

# References

1. Chávez, E., Navarro, G., Baeza-Yates, R., Marroquín, J.L.: Searching in metric spaces. ACM Computing Surveys 33, 273–321 (2001)
2. Zezula, P., Amato, G., Dohnal, V., Batko, M.: Similarity search. The metric space approach. Advances in Database Systems, vol. 32. Springer (2006)
3. Hjaltason, G.R., Samet, H.: Index-driven similarity search in metric spaces. ACM Transactions on Database Systems 28(4), 517–580 (2006)
4. Kalantari, I., McDonald, G.: A data structure and an algorithm for the nearest point problem. IEEE Transactions on Software Engineering 9, 631–634 (1983)
5. Uhlmann, J.K.: Satisfying general proximity/similarity queries with metric trees. Information Processing Letters 40, 175–179 (1991)
6. Brin, S.: Near neighbor search in large metric spaces. In: Procs. of Conf. on Very Large Databases (VLDB 1995), pp. 574–584. Morgan Kaufmann Publishers (1995)
7. Dehne, F., Noltemeier, H.: Voronoi trees and clustering problems. Information Systems 12(2), 171–175 (1987)
8. Navarro, G.: Searching in metric spaces by spatial approximation. In: Procs. of String Processing and Information Retrieval (SPIRE 1999), pp. 141–148. IEEE CS Press (1999)
9. Ciaccia, P., Patella, M., Zezula, P.: M-tree: An efficient access method for similarity search in metric spaces. In: Procs. of Conf. on Very Large Databases (VLDB 1997), pp. 426–435. ACM Press (1997)
10. Traina Jr., C., Traina, A.J.M., Seeger, B., Faloutsos, C.: Slim-Trees: High Performance Metric Trees Minimizing Overlap between Nodes. In: Zaniolo, C., Grust, T., Scholl, M.H., Lockemann, P.C. (eds.) EDBT 2000. LNCS, vol. 1777, pp. 51–65. Springer, Heidelberg (2000)
11. Chávez, E., Navarro, G.: A compact space decomposition for effective metric indexing. Pattern Recognition Letters 26(9), 1363–1376 (2005)
12. Bozkaya, T., Ozsoyoglu, M.: Distance-based indexing for high-dimensional metric spaces. In: Proc. of the ACM Conf. on Management of Data (SIGMOD 1997), pp. 357–368. ACM Press (1997)
13. Novak, D., Batko, M., Zezula, P.: Metric index: An efficient and scalable solution for precise and approximate similarity search. Information Systems 36(4), 721–733 (2009)
14. Skopal, T., Pokorný, J., Snásel, V.: Pm-tree: Pivoting metric tree for similarity search in multimedia databases. In: Procs. of Advances in Database Systems (ADBIS 2004), Local Procs., pp. 803–815 (2004)

# Cut-Region: A Compact Building Block for Hierarchical Metric Indexing

Jakub Lokoč, Přemysl Čech, Jiří Novák, and Tomáš Skopal

SIRET research group, Dept. of Software Engineering, Faculty of
Mathematics and Physics, Charles University in Prague
{lokoc,novak,skopal}@ksi.mff.cuni.cz, Premekcech@seznam.cz

**Abstract.** With the emerging applications dealing with complex multi-media retrieval, such as the multimedia exploration, appropriate indexing structures need to be designed. A formalism for compact metric region description can significantly simplify the design of algorithms for such indexes, thus more complex and efficient metric indexes can be developed. In this paper, we introduce the *cut-regions* that are suitable for compact metric region description and we discuss their basic operations. To demonstrate the power of cut-regions, we redefine the PM-Tree using the cut-region formalism and, moreover, we use the formalism to describe our new improvements of the PM-Tree construction techniques. We have experimentally evaluated that the improved construction techniques lead to query performance originally obtained just using expensive construction techniques. Also in comparison with other metric and spatial access methods, the revisited PM-Tree proved its benefits.

## 1 Introduction

Although there have been many metric access methods (MAMs) [4,19,13,8] developed in the past decades, there still emerge new MAM designs and other approaches addressing the problem of efficient processing of similarity queries. In the last years we observe a trend towards even more complex MAM structures, e.g., M-index [12], pivot table (and all its variants) [11], permutation indexes [5], and others, that are often based on transformation of the metric space into another geometric model. The "good old" hierarchical structures that directly partition the metric space, e.g., M-tree, (m)vp-tree, GNAT, etc., are often outperformed by the new MAMs. From this perspective, it might seem that hierarchical MAMs relying on direct partitioning of the metric space bring an unnecessary overhead and so they should be abandoned. However, although the hierarchical MAMs exhibit worse performance in traditional queries, such as the range query or the $k$ nearest neighbor query, for modern retrieval modalities the hierarchies of metric regions could perform much better. For instance, various iterative queries within the *multimedia exploration* area [2] could benefit from the native hierarchy of metric regions where a continuous traversal in the metric space is required.

In this paper, we define a new formalism for construction of compact metric regions – the *cut-regions*. The cut-regions allow to simplify the development of

G. Navarro and V. Pestov (Eds.): SISAP 2012, LNCS 7404, pp. 85–100, 2012.
© Springer-Verlag Berlin Heidelberg 2012

sophisticated algorithms for construction of a hierarchical MAM. We show how the formalism can be used for re-definition of the PM-tree structure and the construction algorithms. Based on the cut-regions we introduce new PM-tree construction algorithm that leads to more compact PM-tree hierarchies (and so faster similarity search). Note that compact hierarchy of metric regions is not only beneficial for efficiency of traditional queries (range or kNN), but it can also better serve as a hierarchy of clusters that can be used in exploration queries, data mining, and other tasks.

## 1.1 Paper Contributions

The paper contributions can be summarized into three main points:

- The new cut-region formalism, that is suitable for simplified description of compact metric regions. Cut-regions can be utilized in new or existing metric indexing structures and algorithms, as demonstrated on the PM-Tree.
- New cheap dynamic construction techniques for the PM-Tree, that can compete with expensive multi-way leaf selection strategies.
- Thorough experimental evaluation including also comparison with the state-of-the-art MAMs and comparison with the spatial indexing structure R-Tree.

Since the idea of cut-region is the key concept used in this paper, we first define the cut-region and basic operations on them in the following Section 2. Then we recall and redefine the PM-Tree using the cut-regions in Section 3. In Section 4, we present our new dynamic PM-Tree construction techniques and provide thorough experimental evaluation in Section 5. Finally, we conclude the paper and describe our future work in Section 6.

## 2    Cut-Regions

The popular simple ball-regions, defined only by the center object and the covering radius, have one main drawback – they cannot tightly describe a cluster of similar objects in a sparse metric space (see the sparse query ball-region in Figure 1). Moreover, for metric spaces suffering from higher intrinsic dimensionality, the ball-regions become useless. The reason is simple – since only the center object is considered as a pivot, there is no additional information describing relations between the remaining objects in the region. However, if a static set of $k$ global pivots is employed, the remaining objects can be ordered to each pivot separately and the original ball-region can be further cut off by rings (where a ring is an annulus centered in pivot). In other words, the ring for a particular global pivot is determined by the distances to the closest and the farthest object to the pivot. The definition can be further extended to support list of rings for each pivot. An example of the difference between cut-region and the ball-region is depicted in Figure 1. The idea of cut-regions was first used in the PM-Tree [14,18], though here it was not described as a standalone formalism but as a part of the PM-Tree itself. Since cut-regions with their operations can be utilized also in other metric indexes, we have decided to separate this compact metric unit from the PM-Tree into the following definition.

**Definition 1 (Cut-Region).** *Let* $(\mathbb{U}, \delta)$ *be a metric space, $Ball(o, r_o)$ be a subset in a database* $\mathbb{S} \subset \mathbb{U}$ *delimited by a ball* $(o, r_o)$*,* $p_i \in \mathbb{P} \subset \mathbb{U}$ *be k global pivots in a pivot set* $\mathbb{P}$*, and hr set of k intervals where* $hr_i = \langle hr_i^{\min}, hr_i^{\max} \rangle$*,* $hr_i^{\min} = \min_j \{\delta(p_i, o_j)\}$*,* $hr_i^{\max} = \max_j \{\delta(p_i, o_j)\}$*,* $\forall o_j \in Ball(o, r_o)$*, then triplet* $CR(o, r_o, hr)$ *is called the cut-region. If* $r_o = 0$ *then all the intervals hr are set to* $hr_i^{\min} = hr_i^{\max} = \delta(p_i, o)$ *and the cut-region represents only a simple point. Such cut-region is denoted as* $CR(o, 0, hr^o)$*.*

The cut-region in combination with an appropriate set of global pivots is supposed to be a core representation of a metric space region. First, the cut-region allows to determine a center of the cluster $o$ and controls the proximity of the objects via the radius $r_o$. Second, the cut-region utilizes the rings to cut off the "empty space" of the original ball-region. In the task of the metric space clustering and indexing, the cut-region is a suitable unit for a compact cluster description and representation. It is comparable to permutation-based regions, where the proximity of two objects is approximated by the similarity of their permutations, however, the cut-region further controls the object locality via the ball-region center $o$ and radius $r_o$.

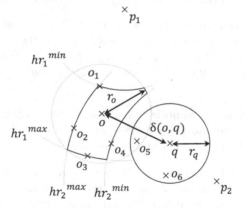

**Fig. 1.** Cut-region and a query ball-region

In the following paragraphs, we propose few definitions and lemmas enabling simpler definition and description of index operations employing cut-regions. The proofs are omitted for the lack of the space, however, they are either trivial or they can be simply derived from lemmas in Section 7 of [19]. For correct filtering during range or kNN query processing, we propose the following lemma.

**Lemma 1.** *Let* $CR = Cut(o, r_o, hr)$ *be a cut-region and* $B = Ball(q, r_q)$ *be a ball-region, if* $\delta(o, q) > r_o + r_q$ *or for some interval* $hr_i$ *it holds* $hr_i \cap \langle \delta(p_i, q) - r_q, \delta(p_i, q) + r_q \rangle = \emptyset$ *then CR and B do not share any object.*

During indexing, new objects could be inserted into existing cut-regions. Since a new object represents a trivial cut-region, the following definition formalizing the inclusion of two cut-regions is proposed. Let us also denote, that the inclusion test (or a weaker form of the test) can be requested for two nontrivial cut-regions ($r > 0$) in hierarchically organized indexing structures.

**Definition 2 (Cut-regions inclusion).** *The cut-region $CR_A = Cut(o_A, r_A, hr^A)$ includes a cut-region $CR_B = Cut(o_B, r_B, hr^B)$ if $\delta(o_A, o_B) + r_B \leq r_A$ and also for each interval $hr_i^B$ holds $hr_i^B \subseteq hr_i^A$ . Formally we write $CR_B \subseteq CR_A$.*

The cut-region inclusion is a geometric relation between two cut-regions $CR_A$ and $CR_B$.

## 2.1   Operations on Cut-Regions

In this subsection we defined two operations, the cut-region extension and the cut-region reduction that are supposed to be frequently employed operations in the dynamic indexing techniques.

At some point of indexing it is impossible to fit the new object into an existing cut-region. Then some suitable cut-region has to be selected and extended by the new inserted object (again treated as the cut-region) or even by a new inserted cut-region in the case of a bulk-loading operation. For such reasons, we define the cut-region extension (the left arrow in Figure 2) as follows.

**Definition 3 (Cut-region extension).** *Let $CR_A = Cut(o_A, r_A, hr^A)$ and $CR_B = Cut(o_B, r_B, hr^B)$ be two cut-regions, the region $CR_E = Cut(o_A, r_E, hr^E)$, where $r_E = \max\{\delta(o_A, o_B) + r_B, r_A\}$ and $hr_i^E = hr_i^A \cup' hr_i^B$, represents the extension of the cut-region $CR_A$ by the cut-region $CR_B$. Formally we write $CR_E = CR_A \leftarrow CR_B$.*

*Note 1.* The cut-region extension is not a commutative operation, because we expect that the first operand $CR_A$ represents an index node, while the second $CR_B$ is always somehow absorbed by the first one.

Since the holes in the intervals are not allowed in our strict cut-region definition, the slightly modified operation $\cup'$ for two intervals, removing potential holes, has to be utilized. Then, the cut-region extension satisfies also the cut-region definition as stated in the following lemma.

**Lemma 2.** *Let $i_1, i_2$ be two intervals, the cut-region extension utilizing operation $\cup'$ for intervals, defined as $i_1 \cup' i_2 = \langle \min\{i_1^{min}, i_2^{min}\}, \max\{i_1^{max}, i_2^{max}\}\rangle$, is cut-region.*

The dynamic rearrangements of the objects within an index can significantly improve the index performance. For such reasons, some objects from a cut-region can be removed from their former locations and reinserted into the new ones and thus we discuss also a cut-region reduction operation (the right arrow in Figure 2). For simplicity, we consider only removal of a trivial cut-region with the zero radius and we also disable removal of $o_A$ from $Cut(o_A, r_A, hr^A)$.

**Definition 4 (Cut-region reduction).** *Let* $CR_A = Cut(o_A, r_A, hr^A)$ *and* $CR_B = Cut(o_B, 0, hr^{o_B})$ *be cut-regions,* $o_A \neq o_B$ *and* $o_B \in Ball(o_A, r_A)$, *then* $CR_R = Cut(o_A, 0, hr^{o_A}) \leftarrow Cut(o_i, 0, hr^{o_i}), \forall o_i \in Ball(o_A, r_A), o_i \neq o_B$, *is called the cut-region reduction. Formally we write* $CR_R = CR_A - CR_B$.

Using $Cut(o_A, 0, hr^{o_A}) \leftarrow Cut(o_i, 0, hr^{o_i}), \forall o_i \in Ball(o_A, r_A), o_i \neq o_B$, the reduction operation creates constructively the reduced cut-region from the scratch. In the following definition we describe vectors that store information useful for indexing algorithms (Figure 3), that need to select cut-regions based on some criteria. The criteria are often based on an aggregation of the vector components.

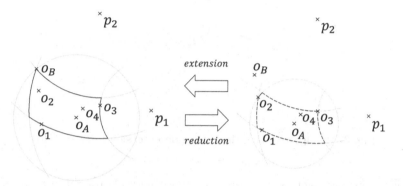

**Fig. 2.** Cut-region reduction

**Definition 5 (Change vector of cut-region extension and reduction).** *Let* $CR_A = Cut(o_A, r_A, hr^A)$ *and* $CR_B = Cut(o_B, r_B, hr^B)$ *be two cut-regions, the* $k + 1$ *dimensional change vector* $cv_e$ *for extension* $CR_A \leftarrow CR_B$ *is defined as* $\{\max(\delta(o_A, o_B) + r_B - r_A, 0), \max(hr_1^{A\,min} - hr_1^{B\,min}, 0) + \max(hr_1^{B\,max} - hr_1^{A\,max}, 0), \ldots, \max(hr_k^{A\,min} - hr_k^{B\,min}, 0) + \max(hr_k^{B\,max} - hr_k^{A\,max}, 0\}$. *If* $CR_B = Cut(o_B, 0, hr^{o_B})$, $o_A \neq o_B$ *and* $o_B \in Ball(o_A, r_A)$, *then the change vector* $cv_r$ *for cut-region reduction* $CR_R = CR_A - CR_B$ *is defined using the change vector for the cut-region extension* $CR_R \leftarrow CR_B$.

In the following section we re-define the PM-Tree using the cut-regions, including both the PM-tree structure and algorithms.

## 3   PM-Tree Revisited

The PM-Tree [14,18] is a metric index that conceptually merges the pivot table [11] with the M-Tree [6]. More specifically, the PM-Tree enhances the original M-tree hierarchy by an information related to the static set of $k$ global pivots $p_i \in \mathbb{P} \subset \mathbb{U}$. Thus, the ground entries (representing data) contain also distances to the global pivots, while the original M-Tree-inherited ball region is further cut off by a set of rings (centered in the global pivots), so the region "volume"

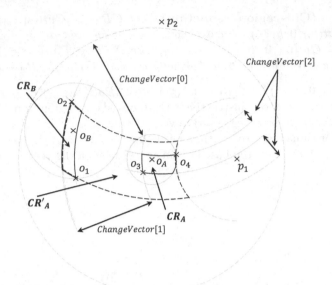

**Fig. 3.** Change vector of the cut-region extension $CR_A \leftarrow CR_B$

becomes more compact (see Figure 4a). If we look thoroughly on the definition of the PM-Tree routing item [14,18], we observe the PM-Tree has already informally introduced (and inspired) the cut-regions.

In the following subsections we quickly review the basic PM-Tree principles and slightly redefine the structure of the PM-Tree routing and ground entries, because we want to adapt the PM-Tree structure to use the cut-region formalism. The data structure representing the cut-region is defined as:

$$CR(y) = [y, r_y, hr^{min}, hr^{max}],$$

where $y$ is the center of the cut-region, $r_y$ is the maximal distance $\delta(y, o_i), \forall o_i \in CR(y)$ and $hr^{min}/hr^{max}$ are $k$-dimensional arrays of (min/max) distances to $k$ global pivots.

### 3.1 PM-Tree Structure

The PM-Tree consists of inner nodes with routing entries and leaf nodes with ground entries. A routing entry in a PM-Tree inner node is defined as:

$$rout_{PM}(y) = [CR(y), \delta(y, \mathrm{Par}(y)), ptr(T(y))],$$

where $CR(y)$ is a cut-region, $\delta(y, \mathrm{Par}(y))$ is the distance from $y$ to the parent routing object, and $ptr(T(y))$ is a pointer to the child node. A ground entry in a PM-Tree leaf node is defined as:

$$grnd_{PM}(z) = [CR(z), id(z), \delta(z, \mathrm{Par}(z))],$$

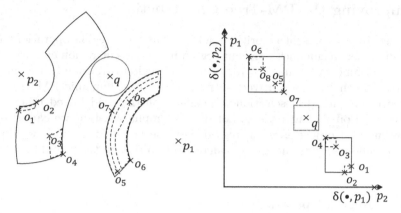

**Fig. 4.** (a) PM-Tree with 2 pivots ($p_1$, $p_2$). (b) Projection of PM-Tree in pivot space.

where $CR(z)$ is a cut-region[1], $\delta(y, \mathrm{Par}(y))$ is the distance from $y$ to the parent routing object, and $id(z)$ is a unique identifier of object $z$.

The combination of all the $k$ entries' ranges (stored in the cut-regions) produces a $k$-dimensional minimum bounding rectangle (MBR), and hence the global pivots actually map the metric regions/data into a pivot space of dimensionality $k$ (see Figure 4b). The number of pivots can be defined separately for routing and ground entries—we typically choose fewer pivots for ground entries to reduce storage cost (i.e., $k = k_{innerNode} > k_{leafNode}$). The pivot space mapping abstraction is much like that one used in pivot tables; however, in the PM-Tree case the pivot space also includes the hierarchy of MBRs (and so resembles R-tree partitioning to some extent).

## 3.2   PM-Tree Querying and Construction

When issuing a range or $k$NN query, the query object is mapped into the global pivot space—this requires $p$ extra distance computations $\delta(q, p_i), \forall p_i \in \mathbb{P}$. The query processing starts in the root node and checks all routing entries' cut-regions for overlap with the query ball applying Lemma 1. If a cut-region cannot be filtered, the child node of the corresponding routing entry is visited. If a leaf node is reached, all stored cut-regions are checked using Lemma 1 and all non-filtered objects are included in the (candidate) result set. Besides the Lemma 1, also the parent filtering rule [6] can be utilized to improve the filtering power of the PM-Tree. Note that applying Lemma 1 does not require many explicit distance computations, so the PM-Tree usually achieves significantly lower query cost when compared with the M-Tree [14,18,15,17].

The PM-Tree construction algorithm is very simple extension of the original M-Tree construction algorithm. The only difference is the maintenance of the $hr^{min}/hr^{max}$ arrays stored in the cut-regions.

---

[1] In the leaf node, the radius is set to zero and $hr^{min}/hr^{max}$ refer to the same array.

# 4   Improving the PM-Tree Construction

In this section, we present algorithms for the basic cut-region operations used almost in all new techniques – the cut-region inclusion, cut-region extension and how to determine the change vector of the cut-region extension (all defined in Section 2). Then we consecutively introduce our new dynamic PM-Tree construction techniques (leaf selection strategies, splitting and forced reinserting) that are based on the cut-regions. Let us also emphasize, that we can consider the new inserted object as a cut-region. For the lack of the space we do not provide pseudo-codes for all the new algorithms, although they would make the techniques more clear.

## 4.1   Cut-Region Operations

The implementation of cut-region operations CUTREGIONINCLUSION($CR_A$, $CR_B$, $numberOfPivots$) or CUTREGIONEXTENSION($CR_A$, $CR_B$, $numberOfPivots$) consists of one simple loop, where the $hr^{min}$ and $hr^{max}$ are employed/updated.

In the Algorithm 4.1 we demonstrate how to determine and aggregate the change vector of the cut-region extension. The result serves as a criterion in the leaf selection strategies or the node split function using cut-regions.

The LOSSOFCUTREGIONREDUCTION($CR_A$, $CR_B$, $numberOfPivots$, $bestResultSoFar$, $agr$) algorithm used for example in the new forced reinserting algorithm can be implemented using CUTREGIONREDUCTION(...) and GROWTHOFCUTREGIONEXTENSION(...) algorithms.

**Algorithm 4.1.** GROWTHOFCUTREGIONEXTENSION(
$CR_A$, $CR_B$, $numberOfPivots$, $bestResultSoFar$, $agr$) $\mapsto$ Result

---

1:  Let $ChVector$ be a (numberOfPivots $+1$) dimensional vector of zeros
2:  **if** $CR_A.r_y < \delta(CR_A.y, CR_B.y) + CR_B.r_y$ **then**
3:      $ChVector[0] = \delta(CR_A.y, CR_B.y) + CR_B.r_y - CR_A.r_y$
4:  **for** ($i = 1$; $i < numberOfPivots + 1$; $i++$) **do**
5:      **if** $CR_A.hr^{min}[i] > CR_B.hr^{min}[i]$ **then**
6:          $ChVector[i]$ += $CR_A.hr^{min}[i] - CR_B.hr^{min}[i]$
7:      **if** $CR_A.hr^{max}[i] < CR_B.hr^{max}[i]$ **then**
8:          $ChVector[i]$ += $CR_B.hr^{max}[i] - CR_A.hr^{max}[i]$

9:      $Result = agr(ChVector)$ // aggregate values of $ChVector$
10:     **if** $Result > bestResultSoFar$ **then** // impossible to get better result
11:         **break**

12: **return** $Result$

---

In this subsection we have introduced basic cut-region algorithms that are used as puzzle pieces in the new PM-Tree construction techniques in the following subsections.

## 4.2   Leaf Selection Strategies Using Cut-Regions

The PM-Tree [18] is generally built in the bottom-up manner so a suitable leaf
selection strategy has a crucial impact on the quality of the resulting hierarchy as
has been shown for its predecessor M-Tree in [10] and [16]. The original PM-Tree
construction technique is very simple – as in the original M-Tree construction
technique, the PM-Tree employs only the ball regions determined by its routing
entries. In fact, the PM-Tree construction algorithms ignore the rings used to
cut empty space in the covering ball regions, thus a newly inserted object can
drastically increase the volume of the corresponding cut-regions. Therefore, our
new leaf selection strategy considers also the rings delimiting the borders of the
cut-regions. In other aspects, the new leaf selection strategy follows the necessary
rules that preserve the original PM-Tree invariants [18].

More specifically, the new technique utilizing the cut-regions changes the notion of the 'good' candidate node for all variants of the leaf selection strategies
(single/multi/hybrid-way). For the single-way leaf selection strategy, such node
becomes the best candidate, the parent routing cut-region of which covers the
newly inserted object (wrapped in the cut-region) and its routing object is as
close to the new object as possible. All covering child nodes of the best candidate
node are then followed down to the next PM-Tree level, while, again, only the
best one is selected as the candidate node, and so on. After the pre-leaf level
is reached, the candidate pre-leaves are checked for the best routing entry and
the respective leaf is returned as the finally selected leaf. In the situation when
no candidate node can be selected at a level (i.e., the new object is not covered
by any node's cut-region), the technique selects such node that guarantees the
minimal growth of its cut-region (for more details see Algorithm 4.2). For the
multi-way leaf selection strategy, more nodes can become good candidates (cut-
regions of their parent routing items cover the newly inserted object) and all
covering child nodes of the candidate nodes are then followed down to the next
PM-tree level. The multi-leaf selection leads to the optimal leaf node, however,
for much higher construction costs. For more details about the multi/hybrid-way
leaf node selection techniques see [16].

## 4.3   Node Splitting Using Cut-Regions

The new node split algorithm is based on the original (P)M-Tree splitting – the
distance matrix is evaluated, two new routing items are selected from the overfull
node and all the remaining entries are distributed between them. Similarly as
in the leaf selection strategy, the original (P)M-Tree split algorithm considering
only ball regions deteriorates the newly created cut-regions in the PM-Tree.
Hence, our new split algorithm focuses on the "volume" of candidate cut-regions.
However, to create the candidate cut-regions and check their properties is much
more time consuming task. Therefore, instead of all possible pairs of candidates
only a fraction of all possible pairs is checked.

When assigning an entry $e_k$ from the overfull node to a candidate routing
item $cr_i$, a growth score function $GS(cr_i, e_k) = $ GrowthOfCutRegionEx-
tension($cr_i$, $e_k$, $numberOfPivots$, $MAX\_Value$, $SUM$) is utilized. Checking

**Algorithm 4.2.** SINGLEWAYFORCUTREGIONS(
$PMTreeInnerNode$, $NewPointCR$, $numberOfPivots$) $\mapsto$ *Leaf*

---

1: $candidate = null$
   //first try to find node containing new region - change vector of the extension $= 0$
2: **for each** $routingEntry$ **in** $PMTreeInnerNode$ **do**
3:    **if** CUTREGIONINCLUSION($routingEntry.CR$, $NewPointCR$, $numberOfPivots$)
      **and** $\delta(routingEntry.CR.y, NewPointCR.y)$ is minimal **then**
4:      $candidate = routingEntry$

5: **if** $candidate$ **is** $null$ **then**     //we haven't found perfect candidate yet
6:    $bestResultSoFar = $ MAX_VALUE
7:    **for each** $routingEntry$ **in** $PMTreeInnerNode$ **do**
8:      $result = $ GROWTHOFCUTREGIONEXTENSION($routingEntry.CR$,
   $NewPointCR$, $numberOfPivots$, $bestResultSoFar$, $SUM$)
9:      **if** $result < bestResultSoFar$ **then**
10:       $candidate = routingEntry$
11:       $bestResultSoFar = result$

12: CUTREGIONEXTENSION($candidate.CR$, $NewPointCR$)

13: **if** $candidate.Ptr(T(y))$ is leaf **then**
14:    **return** $candidate.Ptr(T(y))$

15: **return** SINGLEWAYFORCUTREGIONS($candidate.Ptr(T(y))$,    $NewPointCR$,
   $numberOfPivots$)

---

one candidate pair consists of two steps, both employing the CUTREGIONEX-
TENSION(...) operation. First, until the minimal utilization is reached, the entries $e_k$ with minimal $GS(cr_i, e_k)$ and $GS(cr_j, e_k)$ are alternatively distributed
between the two candidates $cr_i$ and $cr_j$ (cut-region extension operation is performed after each assignment). When the minimal utilization is reached, the
entry $e_k$ is assigned to $cr_i$ if $GS(cr_i, e_k) < GS(cr_j, e_k)$, otherwise the entry $e_k$ is
assigned to $cr_j$.

Finally, we select such candidate pair the greater cut-region of which is minimal among all candidate pairs (analogy to $mMaxRad$ heuristic from the (P)M-
Tree). The size of the cut-region is measured as sum of all ring widths. We denote
this split heuristics as the $mCutReg$.

## 4.4   Reinsertion Using Cut-Regions

When redesigning forced reinsertion for cut-regions, the most important question
is – which objects are optimal for forced reinserting from an overfull node?
Instead of considering only the most distant objects from the parent routing
entry, the $hr^{min}$ and $hr^{max}$ distances to global pivots also have to be taken into
account. Trivially, for $k$ reinserted objects, all possible $k$-tuple candidates can
be checked (i.e., subtracted from the cut-region while the resulting cut-region is

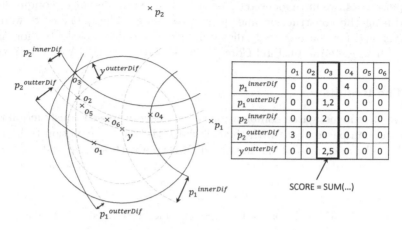

**Fig. 5.** Forced reinsertion from a cut-region

scored). To reduce an indisputable overhead of the trivial solution, we propose a simple heuristics selecting sub-optimal candidate tuple of objects for reinserting. Let $p_i^{innerDif}/p_i^{outerDif}$ be the absolute value of difference of two closest/farthest objects from pivot $p_i$ (see Figure 5). Then, each object $o_j$ closest/farthest from pivot $p_i$ can be assigned a value $p_i^{innerDif}/p_i^{outerDif}$ (see the table in Figure 5). Such value can be utilized as a criterion for reinserting – the value determines how much the cut-region is reduced (according to the pivot $p_i$). Moreover, if an object $o_j$ has assigned more $p_i^{innerDif}/p_i^{outerDif}$ values, their sum determines the overall reduction of the cut-region. Hence, we use this sum as a score function when determining a $k$-tuple of objects that should be reinserted – we reinsert $k$ objects with the highest score. To the resulting score function, we include also $y^{outerDif}$ that represents the reduction of the cut-region radius $r_y$ ($y$ is a local pivot). The subsequent reinsertion processing is the same as used in [10] and [16] for forced reinserts in the M-Tree.

## 5    Experimental Evaluation

### 5.1    The Testbed

We made use of the *ALOI* database [7] comprising 72,000 images extracted in the same way as in [1], the *CoPhIR* subset [3] comprising 641,000 76-dimensional feature vectors (12-dimensional color layout and 64-dimensional color structure), *ColorHistograms* (subset of Corel [9]) comprising 68,000 color histograms, and synthetic dataset *Synthetic* comprising 250,000 10-dimensional points randomly generated from unit hypercube. To compare two feature signatures from the ALOI dataset, the *Signature Quadratic Form Distance* using *Gaussian Similarity Function* was employed (particular alpha is always denoted in graphs). The other datasets were vectorial and the Euclidean distance was employed. In the experiments, we have focused on both real time and distances computations spent during the construction and query processing. The query costs were always averaged for 200 uniformly distributed query objects (only 100 for ALOI dataset). The tests ran on Intel Core i7 920 3.4 GHz, 9 GB RAM, Win 7 x64.

## 5.2   The Results

First, we have investigated combinations of the new proposed construction techniques and compared them with the original PM-Tree construction techniques. Second, we have compared the improved PM-Tree with several state-of-the-art metric and spatial access methods. For better orientation we use only abbreviations for all the methods based on PM-Tree (see Table 1).

**Table 1.** Abbreviations of the presented methods based on PM-Tree

| Method | Description |
|---|---|
| SW_BR | single-way leaf selection using ball regions |
| SW_BR_RI | single-way leaf selection using ball regions and reinserting |
| MW_BR | multi-way leaf selection using ball regions |
| MW_BR_RI | multi-way leaf selection using ball regions and reinserting |
| SW_CR | single-way leaf selection using cut-regions |
| SW_CR_RI | single-way leaf selection using cut-regions and reinserting |
| MW_CR | multi-way leaf selection using cut-regions |
| MW_CR_RI | multi-way leaf selection using cut-regions and reinserting |

All the methods using ball-regions (marked with BR) employ the $mMaxRad$ node splitting strategy, while all methods using cut-regions (marked with CR) employ the $mCutReg$ node splitting strategy described in Section 4. We have set the capacity of the PM-Tree nodes to 20 entries and the minimal utilization of the node to 40%. We have used the same number of pivots both for ground and routing entries.

## 5.3   Comparison with the Original PM-Tree Construction Techniques

First we have focused on the overall comparison of all new proposed construction techniques based on cut-regions with the original PM-Tree construction techniques based on ball-regions. In the top two graphs in Figure 6 the query and indexing costs are presented for the growing number of global pivots. As you can see, the best method for fast query processing is still the original MW_BR_RI, but it has also very high indexing cost. However, for a sufficient number of pivots all the methods based on cut-regions are getting really close to MW_BR_RI query performance, but with significantly lower indexing cost. Especially, cheap SW_CR and SW_CR_RI become promising methods both in

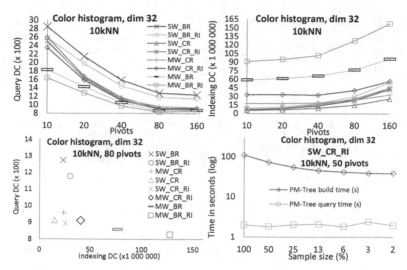

**Fig. 6.** Indexing and querying costs depending on the growing number of global pivots. The top legend is the same for both graphs in the top part. The bottom part shows indexing vs. query cost and the effect of the number of sampled pairs during mCutReg splitting heuristic.

querying and indexing aspects. Let us also emphasize that the cut-region methods based on the expensive multi-way strategy are worse than the cheap single-way approaches.

In the left-bottom graph in Figure 6, a connection between indexing and query cost of all methods is depicted. The original methods (SW_BR, SW_BR_RI, MW_BR, MW_BR_RI) suffer from the classic trade-off problem – lower query cost is paid by expensive indexing, and vice versa. The methods utilizing the cut-regions are exhibit better trade-off – especially SW_CR and SW_CR_RI can be considered as cheap variants of the expensive multi-way techniques. In the right-bottom graph in Figure 6 the effect of the sample size used during *mCutReg* splitting is investigated. As you can see, if all possible pairs are used the indexing is really expensive, while the query time is the same as for much smaller sample sets. Hence, we fixed the sample size to 20% of all possible pairs. With this setting, indexing and querying are relatively balanced. Of course, the results also depend on the selected dataset and on distance employed. When using an expensive metric (such as SQFD), the sample set size can be larger (25%) because the distance matrix evaluation cost becomes dominant during node splitting.

In Figure 7, we have focused on cheap single-way indexing methods employing reinserting (i.e., the practical ones) and compared them on two different datasets under varying number of pivots and database sizes. On both datasets, the query and indexing costs of the SW_CR_RI were lower than for SW_BR_RI. For 20-40 pivots on the ALOI dataset, the query cost of the SW_CR_RI was reduced down to 60% of the SW_BR_RI cost.

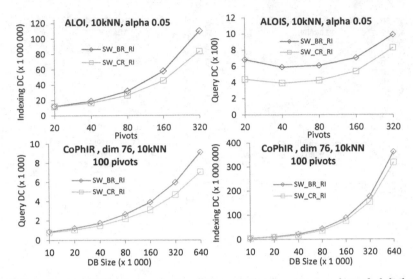

**Fig. 7.** Indexing and querying costs depending on the growing number of global pivots and for varying database sizes

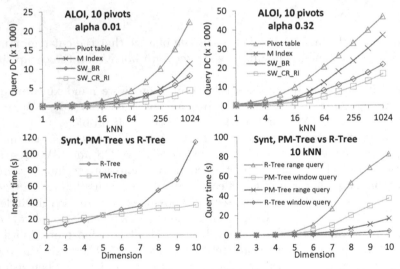

**Fig. 8.** Top part, the comparison with other MAMs. Bottom part, indexing and querying time for PM-Tree and R-Tree depending on growing data dimensionality. Pivot count equals to the corresponding dimension.

## 5.4   Comparison with other Metric and Spatial Indexes

In the last set of experiments, the competitiveness of the improved PM-Tree was validated by inter-MAM comparison and even by comparison with the spatial access method R-Tree. In the top part of Figure 8, we may observe that the

PM-Tree SW_CR_RI outperforms the other MAMs especially for lower query selectivities (i.e., larger query radius). The good performance for larger query radius is caused by the use of many local pivots dynamically created during PM-Tree construction. Hence, the PM-Tree is a good choice for applications where larger query results are expected.

In the bottom part of Figure 8, the PM-Tree and the R-Tree are compared on the low-dimensional synthetic spatial data (the domain of the R-Tree). In the left-bottom graph, we may observe the indexing time, where from dimension five the PM-Tree indexing becomes cheaper. In the right bottom part, both similarity and window queries were investigated, however, only tentatively – we have to emphasize, that the range/kNN query for the R-Tree and the window query for the PM-Tree are not natively supported, and thus were just simply simulated[2]. As expected, the R-Tree works better with window queries, while the PM-tree works better with similarity queries. However, in the case both query types are requested, the PM-Tree can be a good single-index compromise.

## 6   Conclusions and Future Work

In this paper, we have defined the cut-regions enabling construction of compact metric regions. The formalism was used to re-define the PM-Tree and also used to improve the PM-Tree indexing techniques, which resulted in both cheaper and more compact hierarchies. The proposed new techniques are applicable in practice (cheap indexing) and, at the same time, the techniques can even compete in the query performance with order-of-magnitude more expensive multi-way strategies. The PM-Tree also demonstrated its competitiveness in comparison with other indexing structures. In the inter-MAM comparison (including Pivot Tables and M-Index) the PM-Tree proved best performance for queries with larger radius. Finally, the PM-Tree can be used as a spatial access method providing good performance for similarity queries and not so bad performance for window queries. In the future, we plan to investigate batch loading strategies using cut-regions, that will make the PM-Tree indexing even cheaper. We also plan a more sophisticated window query algorithm in the PM-Tree.

**Acknowledgments.** This research has been supported in part by Czech Science Foundation projects P202/11/0968 and P202/12/P297.

## References

1. Beecks, C., Lokoč, J., Seidl, T., Skopal, T.: Indexing the signature quadratic form distance for efficient content-based multimedia retrieval. In: Proc. ACM International Conference on Multimedia Retrieval (2011)

---

[2] The hypercube representing window query was wrapped into the ball region centered in the center of the hypercube and the ball-region representing similarity query was wrapped by the tightest possible hypercube bounding the ball-region.

2. Beecks, C., Skopal, T., Schoeffmann, K., Seidl, T.: Towards large-scale multimedia exploration. In: Das, G., Hsristidis, V., Ilyas, I. (eds.) Proceedings of the 5th International Workshop on Ranking in Databases (DBRank 2011), pp. 31–33. VLDB, Seattle, WA, USA (2011)
3. Bolettieri, P., Esuli, A., Falchi, F., Lucchese, C., Perego, R., Piccioli, T., Rabitti, F.: CoPhIR: a test collection for content-based image retrieval. CoRR, abs/0905.4627v2 (2009)
4. Chávez, E., Navarro, G., Baeza-Yates, R., Marroquín, J.L.: Searching in metric spaces. ACM Computing Surveys 33(3), 273–321 (2001)
5. Gonzalez, E.C., Figueroa, K., Navarro, G.: Effective proximity retrieval by ordering permutations. IEEE Trans. Pattern Anal. Mach. Intell. 30(9), 1647–1658 (2008)
6. Ciaccia, P., Patella, M., Zezula, P.: M-tree: An Efficient Access Method for Similarity Search in Metric Spaces. In: VLDB 1997, pp. 426–435 (1997)
7. Geusebroek, J.-M., Burghouts, G.J., Smeulders, A.W.M.: The Amsterdam Library of Object Images. International Journal of Computer Vision 61(1), 103–112 (2005)
8. Hetland, M.L.: The Basic Principles of Metric Indexing. In: Coello, C.A.C., Dehuri, S., Ghosh, S. (eds.) Swarm Intelligence for Multi-objective Problems in Data Mining. SCI, vol. 242, pp. 199–232. Springer, Heidelberg (2009)
9. Hettich, S., Bay, S.D.: The UCI KDD archive (1999), http://kdd.ics.uci.edu
10. Jakub, L., Tomáš, S.: On Reinsertions in M-tree. In: SISAP 2008: Proceedings of the First International Workshop on Similarity Search and Applications (SISAP 2008), pp. 121–128. IEEE Computer Society, Washington, DC (2008)
11. Mico, M.L., Oncina, J., Vidal, E.: A new version of the nearest-neighbour approximating and eliminating search algorithm (aesa) with linear preprocessing time and memory requirements. Pattern Recogn. Lett. 15(1), 9–17 (1994)
12. Novak, D., Batko, M., Zezula, P.: Metric index: An efficient and scalable solution for precise and approximate similarity search. Inf. Syst. 36(4), 721–733 (2011)
13. Samet, H.: Foundations of Multidimensional and Metric Data Structures. Morgan Kaufmann (2006)
14. Skopal, T.: Pivoting M-tree: A Metric Access Method for Efficient Similarity Search. In: Proceedings of the 4th Annual Workshop DATESO, Desná, Czech Republic, pp. 21–31 (2004) ISBN 80-248-0457-3; also available at CEUR, vol. 98, ISSN 1613-0073, http://www.ceur-ws.org/Vol-98
15. Skopal, T.: Unified framework for fast exact and approximate search in dissimilarity spaces. ACM Transactions on Database Systems 32(4), 1–46 (2007)
16. Skopal, T., Lokoč, J.: New Dynamic Construction Techniques for M-tree. Journal of Discrete Algorithms 7(1), 62–77 (2009)
17. Skopal, T., Lokoč, J.: Answering Metric Skyline Queries by PM-tree. In: Proceedings of the Dateso 2010 Workshop, vol. 567, pp. 22–37. Matfyz Press (2010)
18. Skopal, T., Pokorný, J., Snášel, V.: Nearest Neighbours Search Using the PM-Tree. In: Zhou, L.-z., Ooi, B.-C., Meng, X. (eds.) DASFAA 2005. LNCS, vol. 3453, pp. 803–815. Springer, Heidelberg (2005)
19. Zezula, P., Amato, G., Dohnal, V., Batko, M.: Similarity Search: The Metric Space Approach. Springer (2005)

# Static-to-Dynamic Transformation
# for Metric Indexing Structures

Bilegsaikhan Naidan and Magnus Lie Hetland

Department of Computer and Information Science,
Norwegian University of Science and Technology,
Sem Sælands vei 7-9, NO-7491 Trondheim, Norway
{bileg,mlh}@idi.ntnu.no

**Abstract.** In this paper, we study the well-known algorithm of Bentley and Saxe in the context of similarity search in metric spaces. We apply the algorithm to existing static metric index structures, obtaining dynamic ones. We show that the overhead of the Bentley-Saxe method is quite low, and because it facilitates the dynamic use of any state-of-the-art static index method, we can achieve results comparable to, or even surpassing, existing dynamic methods. Another important contribution of our approach is that it is very simple—an important practical consideration. Rather than dealing with the complexities of dynamic tree structures, for example, the core index can be built statically, with full knowledge of its data set.

## 1   Introduction

Many modern applications require efficient similarity retrieval, including applications in multimedia (to find similar images, audio in digital-repositories), pattern recognition (to identify finger-prints, face images in image databases), and string searching (to find words in a dictionary while permitting spelling errors). In such applications, the search problem is often stated in terms of distance-search in a metric space. That is, given a metric $d$ over a universe $\mathbb{U}$, and a data set $\mathbb{D} \subset \mathbb{U}$, find the objects in $\mathbb{D}$ that are closest to some query $q \in \mathbb{D}$ (either all within a search radius $r$, or the $k$ nearest neighbors, $k$-NN).

Rather than performing a linear scan of the full data set, it is common to preprocess the data set by building an index structure by exploiting the metric axioms (the triangular inequality in particular). Most existing such index structures are *static*.[1] That is, the index is built with access to the full data set, and if an object is to be added or deleted, a full rebuild of the entire index is required. This is, of course, time consuming and computationally intensive. To accommodate these operations, some special-purpose *dynamic* index structures, supporting additions and deletions at low cost, have been proposed. Maintaining

---

[1] Based on an analysis of the proceedings of the International Workshop on Similarity Search and Applications.

G. Navarro and V. Pestov (Eds.): SISAP 2012, LNCS 7404, pp. 101–115, 2012.

the integrity and performance of such a structure over time, with only incremental information, can be challenging; such structures can be more complicated, as well as less able to utilize global information about the data set.

In this paper, we study the Bentley-Saxe (BS) [1] algorithm in the context of similarity search in metric spaces. BS is a tool that allows us to transform a static data structure into a dynamic one for any decomposable search problem (as explained in Section 3). This means that we can still use the state of the art in static indexing, even if we need the functionality of a dynamic indexing method, without losing the ability to globally analyze the data set, and without adding any appreciable complexity. In fact, the Bentley-Saxe method can use the indexing methods as black-box modules, permitting a clean separation of the (static) indexing and the dynamism.

This paper is organized as follows. Section 2 describes some related work. The Bentley-Saxe method is explained in Section 3. Section 4 provides our experimental results. Some concluding remarks are given in section 5.

## 2    Related Work

In this section we briefly overview some relevant static and dynamic metric indexing structures. For further details, refer to the tutorial by Hetland [7] and book by Zezula et al. [13]. We consider two well-known static methods (the VP-tree and the SSS-tree) as well as two dynamic ones (EGNAT and the DSA-tree).

The vantage point (VP) tree [12] is a static balanced binary tree. During construction, the VP-tree first selects a representative object $p$ (a so-called *vantage point*) from the dataset $\mathbb{D}$ and computes the median $m$ of the distances between $p$ and the other objects in the dataset. Then it divides the dataset into two subsets $\mathbb{D}_1$ and $\mathbb{D}_2$ such that $\mathbb{D}_1 = \{x \in \mathbb{D} \mid x \neq p, d(p, x) \leq m\}$ and $\mathbb{D}_2 = \mathbb{D} \setminus (\mathbb{D}_1 \cup \{p\})$. The algorithm recursively builds left and right subtrees for $\mathbb{D}_1$ and $\mathbb{D}_2$, if they are not empty. A range query $q$ with radius $r$ is performed by recursively traversing the tree from the root to leaves. For each visited node, $d(q, p)$ is computed and $p$ is reported if $d(q, p) \leq r$. It is necessary to visit in the left subtree only if $d(q, p) - r \leq m$, and similarly, for the right subtree only if $d(q, p) + r \geq m$.

There exists a dynamic version of the VP-tree [6]. However, it is not at all straightforward to implement correctly, and in some cases it is still unable to avoid periodic reconstruction of subtrees or even of the entire tree.

Brisaboa et al. have proposed a static index structure so called the Sparse Spatial Selection (SSS) tree [3], in which the first object in a dataset is selected as the first cluster center and then the rest of the objects become new cluster centers if they are far enough away (i.e., the minimum distance between the object and current cluster centers is greater than $\alpha M$, where $\alpha$ is an user-defined parameter and $M$ is the maximum distance between any two objects) from all current centers; otherwise, they are assigned to the cluster associated with the nearest center. The process is recursively applied to those clusters that have not yet fallen below a given size threshold.

The Geometric Near-neighbor Access Tree (GNAT) [2] is a multiway static tree and is built as follows. First, a set of pivots are selected at random and then the rest of the objects are assigned to a region associated with the closest pivot. Examples of a GNAT are shown in Figure 1a, 1b. For each region, the minimum and maximum distances to the other regions' objects are kept for efficiently filtering out non-promising regions in the search, meaning that a region is discarded if the query ball does not intersects with this distance interval. The subtrees are recursively built for all regions associated with the pivots.

The Evolutionary GNAT (EGNAT) [11] is a dynamic version of GNAT. The root is initially created as a leaf node. The insertion algorithm traverses the index structure by choosing the subtree associated with the closest pivot until a leaf node is reached. If the leaf node has a room for the new object, it is added there. Otherwise, the leaf node is transformed into an internal node by selecting pivots and distributing its objects into new child (leaf) nodes. The leaf nodes also keep information about distances to their parent objects. During the search this information is used to establish lower bounds to the actual distances between the query and objects.

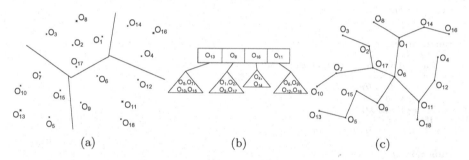

(a)                              (b)                              (c)

**Fig. 1.** Examples of (a) a GNAT space decomposition with hyperplanes between $O_8$, $O_{11}$, $O_{13}$ and $O_{16}$, (b) the corresponding GNAT tree, and (c) a SA-tree with the root $O_6$

The spatial approximation (SA) tree [8] is based on an approach that is completely different from the hierarchical space decomposition of the other trees. First an arbitrary object is selected as the root of the tree and a set of its neighbors is selected as follows. An object is inserted in the neighbor list if it is closer to the root than all current neighbors. Otherwise, the object is assigned to a subset associated with its closest neighbor. Then, for each subset the procedure is applied recursively. Figure 1c shows an example of a SA-tree. The search algorithm uses a best-first branch-and-bound approach, like most metric search algorithms.

Navarro et al. [9] have shown that the SA-tree can be built dynamically, and they call the resulting structure the dynamic SA (DSA) tree. They manage to preserve the semantics of the SA-tree by introducing a *time-stamp* for every object. These time-stamps are then used during search, to ensure that only distance relationships that were known at the time of insertion are used when filtering out objects, to avoid false dismissals.

## 3    The Bentley and Saxe Algorithm

The main data structure of BS is a set of $m = \lfloor \log_2 n \rfloor + 1$ buckets[2] $B_0, B_1, ...,$ $B_{m-1}$ and each bucket $B_i$ is either empty or a static data structure that contains a collection of objects with size $2^i$. To insert a new object into the index, the algorithm follows the same principle that is used for incrementing a binary counter, where the $i$-th bit denotes the absence or presence of a static index structure in the bucket $B_i$. The search is performed by accessing non-empty buckets and combining the results. The pseudo-code is sketched in Algorithm 1. We note that the search starts from $B_{m-1}$ to $B_0$. This may affect to the efficiency of $k$-NN search (by shrinking the covering radius of the current $k$-NN candidate set as much as possible early on).

---

**Algorithm 1.** Static to dynamic transformation

```
1: function Init():
2:     B₀ ← null; m = 0

3: function Insert(x):
4:     D ← {x}
5:     Find minimum k such that Bₖ = null
6:     for i ← 0 to k − 1:
7:         D ← D ∪ Unbuild(Bᵢ)
8:         Bᵢ ← null
9:     Bₖ ← Build(D)
10:    if k = m:
11:        Bₘ₊₁ ← null; m ← m + 1

12: function Query(q):
13:    ans ← ∅
14:    for i ← m − 1 downto 0:
15:        if Bᵢ ≠ null:
16:            Search using q in Bᵢ and update ans with results
17:    return ans
```

---

Let us consider an example of the insertion of a new object into the existing data structure. The example is illustrated in Figure 2.

Let the buckets $B_0, B_1, \ldots, B_{k+2}$ be non-empty. Thus, the first empty bucket is $B_{k+3}$. We build an index structure for bucket $B_{k+3}$ containing the new object and all the objects stored in buckets $B_0, B_1, \ldots, B_{k+2}$. After building this structure, buckets $B_0, B_1, \ldots, B_{k+2}$ are nulled. Buckets $B_{k+4}$ and upward are unchanged.

Now consider the asymptotic running time and space requirements of this approach. Let $T$ be a static metric index structure with size $S_T(n)$ that can be

---

[2] For a dynamic index, the number of buckets is, of course, unknown at the outset. The problem size $n$ is the number of objects added *so far*.

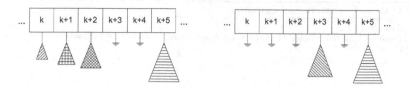

**Fig. 2.** Illustrations of an index structure before the insertion of a new object (left) and after the insertion (right)

built in time $C_T(n)$ and perform a query in time $Q_T(n)$. Under some natural assumptions[3] BS gives us a dynamic metric index structure $T'$ based on $T$ that requires the storage $S_{T'}(n) = \mathcal{O}(S_T(n))$, construction time $C_{T'}(n) = \mathcal{O}(C_T(n))$, query time $Q_{T'}(n) = \mathcal{O}(\log n \cdot Q_T(n))$ and an amortized insertion time for $n$ elements of $I_{T'}(n) = \mathcal{O}(\log n \cdot C_T(n)/n)$.

The original version of BS method was not designed to handle deletions efficiently. Consider, for example, the scenario where we have a single non-empty bucket $B_k$, containing $2^k$ objects. To delete an object now, we have to split $B_k$ into $B_0, B_1, \ldots, B_{k-1}$. This entails building $k$ index structures, might be computationally expensive. To address this, Overmars et al. [10] weakened the condition of the BS method so that every bucket $B_k$ can be either empty or a static data structure which stores at least $2^{k-2}$ and at most $2^k$ objects. With this new condition, our deletion would affect only on $B_{k-2}$, $B_{k-1}$ and $B_k$. The approach of Overmars et al. is shown in Algorithm 2.

In line 7, we mark $o$ as deleted in $B_k$. The bucket $B_k$ might not be rebuilt until its total number of objects becomes $2^{k-2}$. That would, of course, affect the search performance. In order to decrease this effect, we introduce parameter tuning option in between lines 8 and 9. There are a lot of possibilities for the parameter tuning. For instance, the bucket $B_k$ is rebuilt each time when $2^{k-3}$ objects have been deleted from that bucket. This strategy will be tested in our experiments.

## 4   Experiments

In this section we present our experimental evaluation of two new dynamic trees based on BS, comparing them against two existing dynamic trees. As the performance measure we used the number of distance computations required to construct index structures and to answer similarity queries. We have also investigated the overhead of the BS method, by comparing the build and search times of the static indexes to those of their transformed, dynamic counterparts. We have provided performance comparisons of range and $k$-NN queries, as well as deletion costs per object and search performances after deletions.

---

[3] For example, that $S_T$ and $C_T$ are polynomial and that $Q_T$ is at worst linear.

**Algorithm 2.** Overmars and Leeuwen

```
1: function Insert(x):
2:     Replace line 9 of Insert function of Algorithm 1 with the following
           if |D| > 2^{k-1}: B_k ← Build(D)
           else: B_{k-1} ← Build(D)                                    ▷ |D| > 2^{k-2}

3: function Remove(o):
4:     Perform a range search in B_k to find k such that o ∈ B_k      ▷ k from m − 1 downto 0
5:     if not found o:
6:         return false
7:     Delete o from B_k                                              ▷ |B_k| is decremented by 1
8:     if |B_k| > 2^{k-2}:
9:         return true
10:    elif |B_k| = 2^{k-2} and k ≥ 2:
11:        if B_{k-1} ≠ null:
12:            D ← Unbuild(B_k) ∪ Unbuild(B_{k-1})
13:            if |B_{k-1}| > 2^{k-2}:
14:                B_{k-1} ← null
15:                B_k ← Build(D)
16:            else:
17:                B_k ← null
18:                B_{k-1} ← Build(D)
19:        elif B_{k-1} = null and B_{k-2} ≠ null:
20:            D ← Unbuild(B_k) ∪ Unbuild(B_{k-2})                     ▷ |B_k| + |B_{k-2}| > 2^{k-2}
21:            B_k ← null; B_{k-2} ← null
22:            B_{k-1} ← Build(D)
23:        elif B_{k-1} = null and B_{k-2} = null:
24:            D ← Unbuild(B_k); B_k ← null
25:            B_{k-2} ← Build(D)
26:        return true
```

## 4.1  The Testbed

We performed experiments using both synthetic data sets, generated by us, and real-world datasets obtained from the SISAP metric space library [5]. For all vectors we use the Euclidean distance, unless otherwise stated.

- Uniform 10: Synthetic. 100 000 uniformly generated 10-dimensional vectors (synthetic).
- Clusters 10: Synthetic. 100 000 clustered 10-dimensional vectors with 10 cluster centers. The centers were randomly chosen from a uniform distribution and objects in the clusters were generated from the multivariate normal distribution around each of the cluster centers with a variance of 0.1.
- Uniform 20: Synthetic. 100 000 uniformly generated 20-dimensional vectors.
- Clusters 20: Synthetic. 100 000 clustered 20-dimensional vectors with 100 cluster centers. We followed the same procedure as in Clusters 10 to generate the cluster centers.
- NASA: 40 150 feature vectors with 20 dimensions extracted from NASA images.
- Dictionary: a dictionary of 69 069 English words. We use the edit distance (or Levenstein distance), that is, the minimum number of insertions, deletions, and substitutions needed to transform one string into another.
- Histogram: a collection of 112 682 color histograms (112-dimensional vectors) from an image database.

**Table 1.** *idims* of the datasets

| Uniform 10 | Clusters 10 | Uniform 20 | Clusters 20 | NASA | Dictionary | Histogram |
|:---:|:---:|:---:|:---:|:---:|:---:|:---:|
| 13.36 | 9.24 | 27.64 | 20.44 | 5.18 | 8.49 | 2.74 |

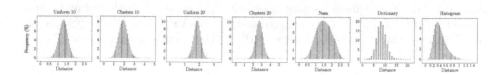

**Fig. 3.** Distance distribution histograms

Table 1 shows the intrinsic dimensionalities (*idims*) [4] of the datasets. The distance histograms of the data sets are shown in Figure 3.

## 4.2 Experiment Settings

We have applied the BS method to VP- and SSS-trees and call the resulting dynamic structures the BS-VP-tree and BS-SSS-tree, respectively. We have compared their performances to two dynamic metric index structures, the DSA-tree and EGNAT. We set the maximum node fanout of the BS-SSS-tree to 5, 10, 20, 40 and 80. The parameter $\alpha$ was 0.45 for the 20-dimensional and 0.40 for the remaining of the datasets. The value of $M$ is estimated before every (re)construction of bucket as follows. An arbitrary object in the bucket is selected as the boundary object. Then, the distances between the boundary and all objects in the bucket are computed 10 times by maximizing the value of $M$ and renewing the boundary object from current one. The cost of this estimation is also included in the construction and deletion costs. We used the SISAP implementation [5] of DSA-tree with time-stamping and bounded arity. The original authors [9, §5.8] suggested this version of DSA-tree that would give the best results in terms of construction cost and search efficiency. The maximum arities of DSA-tree were set to 2, 4, 8, 16 and 32, as in their experiments. For EGNAT, we set the parameters by trial and error. We used internal node sizes of 4, 8, 12, 16 and 20 and maximum leaf node arities of 5, 10, 20, 40 and 80. In total, we performed 18 (5+4+3*3) runs (with several) queries.[4]

We randomly shuffled the order of all objects in each dataset 10 times, obtaining 10 versions of the dataset and the results were averaged over 10 runs using these versions. For each run, a query set consists of 1000 queries which were selected from the respective dataset and the remaining objects in the dataset used for indexing. We selected search radii for range queries so that we capture on average 0.01 %, 0.1 % and 1 % of the vectors. The search radii were in

---

[4] Note that leaf node size should be greater than or equal to internal node size.

the range from 1 to 4 for the dictionary, capturing on average 0.003 %, 0.042 %, 0.361 %, and 1.946 % of the dataset, respectively. For $k$-NN search, we compared the search efficiency of four structures by varying the result size thresholds, using the values 1, 5, 10, 20, 40 and 80. We report only best results in terms of search efficiency from the results obtained with different parameters use on every query set.

For deletions, the deletion costs include both costs of locating the object that to be deleted (this is usually done with range search with radius 0) and rebuilding of buckets. First, we constructed the BS-VP-tree and BS-SSS-tree on the datasets. Then, we deleted every 10 % of the corresponding datasets from BS indexes and obtained the ratio between the number of distance computations required to answer a query set in the *deleted* BS indexes and in *newly built* BS indexes for the remaining objects in the datasets after deletions.[5]

We have implemented our experimental framework in C++, which was compiled in gcc 4.6.2 with the option -O3. All experiments are performed on a PC with a 3.3 GHz Intel Core i5-2500 processor and 8 GiB RAM. We did not use any caching of distances during the construction of index and query processing.

## 4.3    The Overhead of BS Index

First, let us consider the *construction cost* overhead of BS-based index structures. We constructed the VP-tree, SSS-tree (with maximum node fanout 5), BS-VP-tree and BS-SSS-tree (with maximum node fanout 5) 10 times on every 10 % of various datasets and obtained the ratio between the number of distance computations that required to build the BS index and static index with the same settings, with the BS index built incrementally. The mean of ratio was 3.23 (standard deviation 0.61), with minimum and maximum values of 1.69 and 4.27. So in our experiments, the BS index is at most 4.27 times as costly to build as the corresponding static index. Figure 4 shows the construction cost overhead of the BS-based index structures using the box plots that display the minimum, the 25 % quantile, the median, the 75 % quantile and the maximum value.

The figures show that the average ratio for VP-tree is much higher than the SSS-tree, and the values for VP-tree are distributed almost evenly. The values for SSS-tree are positively skewed in general, i.e., it has relatively few high values, and performed well on the dictionary.

Now let us consider the ratio for *index construction time* and *query set execution time* with all $k$ and search radii. We followed the same principle previously used for the construction cost ratio to obtain these ratios with $2^m - 1$ objects for each dataset. The reason is that all the buckets in BS will be non-empty so we intentionally increase the overhead of BS-based index structures, intending to elicit the worst-case search performance. The ratios are shown in Figure 5. For the construction time ratio, the maximum value for BS-VP-tree was 3.62 (with

---

[5] The experiments of deletions with DSA-tree were not performed due to a bug in the SISAP metric space library. We contacted one of the original authors of DSA-tree and it became clear that the bug can not be fixed before the conference's deadline.

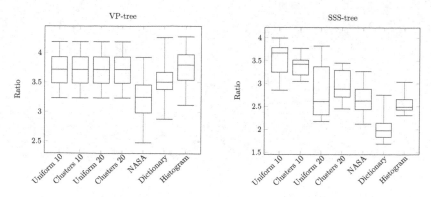

**Fig. 4.** Construction cost ratio of BS index to static index with same settings

construction time 1.23 s) on the dictionary (Figure 5a) while the maximum value for BS-SSS-tree was 3.80 (with construction time 11.09 s) on Uniform 20 (Figure 5d). In Figure 5b, 5c, 5e and 5f, we see that there is almost no search time difference between static and BS-based index structures on the 20-dimensional synthetic vectors due to high *idim*s. In all of our experiments, the maximum value of query set execution time for the static index structures was 19.87 s while for the BS-based index structures the maximum value was 20.97 s.

## 4.4   Comparison of Construction Costs

All index structures were built in an incremental fashion, i.e., initially all of the index structures were empty and then all objects in the datasets were added into the index one by one. The construction costs are shown in Table 2.

**Table 2.** Construction costs of index structures on various datasets

| Dataset | BS-VP-tree | BS-SSS-tree | EGNAT | DSA-tree |
|---------|-----------|-------------|-------|----------|
| Uniform 10 | 5762240 | 59199773 | 3451333 | 3126671 |
| Clusters 10 | 5762240 | 38317600 | 5327960 | 3298234 |
| Uniform 20 | 5762240 | 208912258 | 7582301 | 8631551 |
| Clusters 20 | 5762240 | 96468450 | 8876612 | 8857394 |
| NASA | 1952946 | 13619821 | 1646678 | 1219949 |
| Dictionary | 4676487 | 90720799 | 4820213 | 5531384 |
| Histogram | 6246315 | 45558887 | 9688228 | 4124772 |

As the *idim* of the synthetic datasets increases we see that the datasets become difficult to index. This increase clearly affects the construction cost of the SSS-tree. This effect may be due to the fact that every object tends to become a

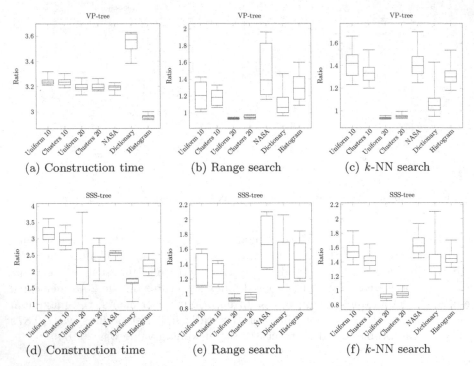

**Fig. 5.** Construction time and query set execution time ratio of BS index to static index with same settings

cluster center of the tree because all objects are approximately equidistant from each other in high-dimensional spaces. It should also be noted that the clustering cost of the SSS-tree is high also in the static case, so this is not an artifact of our approach.

When considering any overhead in construction, it is important to note that our method is quite amenable to *bulk loading*: If a given data set is available at the outset, or if a large number of objects are added, there is *no need* to build the structure incrementally, by adding individual objects. Instead, which buckets need to be filled can be easily calculated from the total data size, and the objects can be partitioned among these (e.g., randomly), and the static structures built. This would mean that there would be no need for multiple rebuilds, and the overhead would be much lower. For example, if the data size were a power of 2, there would be *no overhead whatsoever*. The resulting data structure would still retain all its dynamic properties. (The overhead in general will, of course, vary with how close the data size is to a power of 2, either above or below.)

### 4.5   Query Performance, with and without Deletions

Figure 6 shows the search results over the synthetic datasets and the impact of dimensionality. The performance of all of the index structures degrades when

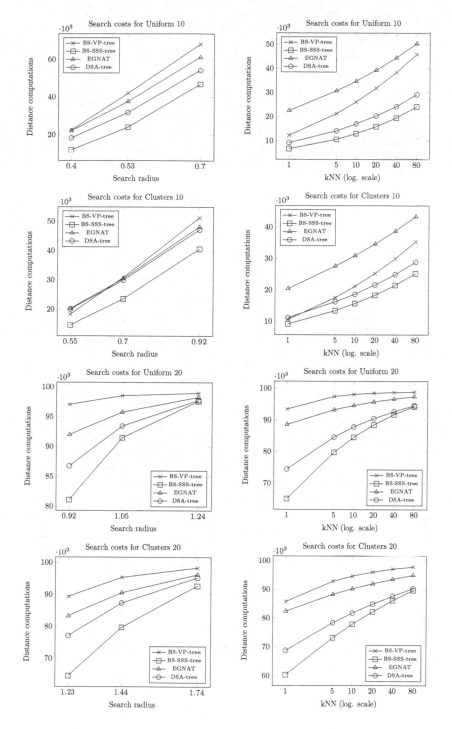

**Fig. 6.** Performance evaluations on the synthetic datasets

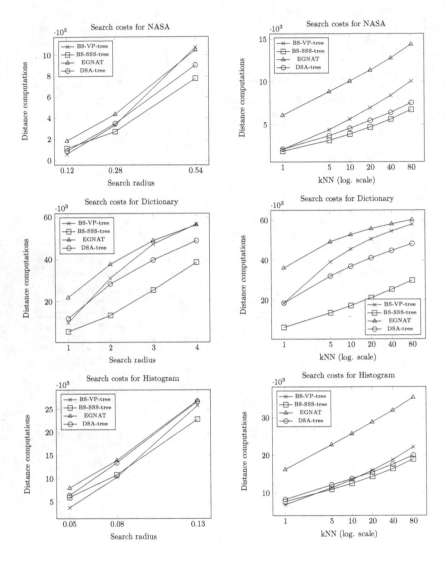

**Fig. 7.** Performance evaluations on the real-world datasets

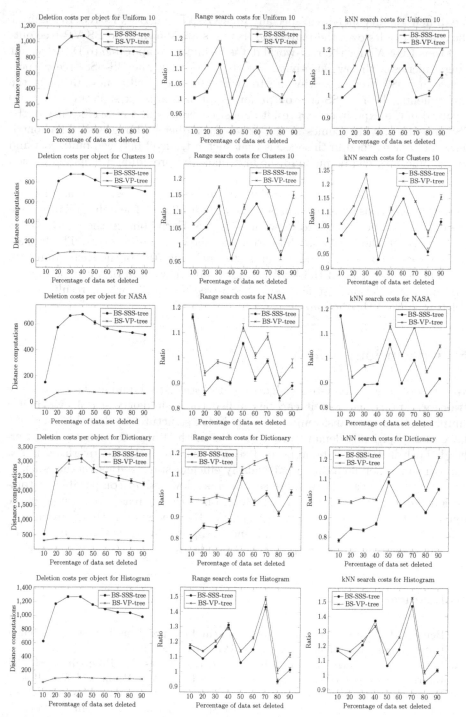

**Fig. 8.** Deletion costs and distance computation ratios of *deleted* BS index to *newly built* BS index

the dimensionality increases, specially for EGNAT and BS-VP-tree. In Figure 7 the search results over the real-world datasets are shown. The BS-VP-tree outperforms EGNAT for the real-world datasets and is comparable to DSA-tree for range queries with low selectivity. For the dictionary, the BS-SSS-tree outperforms the DSA-tree, achieving up to twice the search efficiency. In all of the experiments, the BS-SSS-tree outperforms all of the index structures that are involved in the experiments, a result almost certainly due to the efficiency of the SSS-tree itself, which comes at the price of a higher building cost. The contribution of our method in this case is that such a tradeoff between build cost and search efficiency can be made in the first place, by providing a dynamic version of the SSS-tree.

Deletions were performed on various datasets. We measured the all distance computation ratios (explained in Section 4.2) for all $k$ and search radii over various data sets. The results are shown in Figure 8. Each point in the figures shows the average of those ratios, while error bar shows standard error. The highest ratio of search costs is 1.52, and occurs after deleting 70 % of the Histogram data set. The deletion cost of the BS-VP-tree is quite low in all of our experiments. The highest deletion cost for the BS-SSS-tree was 3129, which resulted from deleting 40 % of the dictionary.

## 5  Conclusions

We studied the Bentley-Saxe algorithm for static-to-dynamic data structure transformations and how it can be applied to in similarity search, yielding a simple method for transforming of static index structures to dynamic ones. We have also empirically demonstrated that the method has a reasonably low overhead, both in terms of building and search cost. In fact, this overhead is low enough that when adapting a particularly efficient static data structure such as the SSS-tree, we can still achieve search times lower than comparable custom-designed dynamic data structures. In addition to this increased performance, the dynamic structures resulting from using the Bentley-Saxe method can be considerably simpler than other dynamic indexes, given that it is simply an isolated add-on to existing (usually simpler) static indexes.

## References

1. Bentley, J.L., Saxe, J.B.: Decomposable searching problems I. Static-to-dynamic transformation. Journal of Algorithms 1(4), 301–358 (1980)
2. Brin, S.: Near neighbor search in large metric spaces. In: Proceedings of 21th International Conference on Very Large Data Bases, VLDB, pp. 574–584 (1995)
3. Brisaboa, N.R., Pedreira, O., Seco, D., Solar, R., Uribe, R.: Clustering-Based Similarity Search in Metric Spaces with Sparse Spatial Centers. In: Geffert, V., Karhumäki, J., Bertoni, A., Preneel, B., Návrat, P., Bieliková, M. (eds.) SOFSEM 2008. LNCS, vol. 4910, pp. 186–197. Springer, Heidelberg (2008)

4. Chávez, E., Navarro, G.: A Probabilistic Spell for the Curse of Dimensionality. In: Buchsbaum, A.L., Snoeyink, J. (eds.) ALENEX 2001. LNCS, vol. 2153, pp. 147–160. Springer, Heidelberg (2001)

5. Figueroa, K., Navarro, G., Chavez, E.: Metric spaces library (2010), http://www.sisap.org/Metric_Space_Library.html (downloaded November 15, 2011)

6. Fu, A.W.-C., Chan, P.M.-S., Cheung, Y.-L., Moon, Y.S.: Dynamic vp-tree indexing for $n$-nearest neighbor search given pair-wise distances. The VLDB Journal 9(2), 154–173 (2000)

7. Hetland, M.L.: The Basic Principles of Metric Indexing. In: Coello, C.A.C., Dehuri, S., Ghosh, S. (eds.) Swarm Intelligence for Multi-objective Problems in Data Mining. SCI, vol. 242, pp. 199–232. Springer, Heidelberg (2009)

8. Navarro, G.: Searching in metric spaces by spatial approximation. The VLDB Journal 11(1), 28–46 (2002)

9. Navarro, G., Reyes, N.: Dynamic spatial approximation trees. Journal of Experimental Algorithmics, JEA, 12:1.5:1–1.5:68 (2008)

10. Overmars, M., Leeuwen, J.: Two general methods for dynamizing decomposable searching problems. Computing 26, 155–166 (1981)

11. Uribe, R., Navarro, G., Barrientos, R.J., Marín, M.: An Index Data Structure for Searching in Metric Space Databases. In: Alexandrov, V.N., van Albada, G.D., Sloot, P.M.A., Dongarra, J. (eds.) ICCS 2006. LNCS, vol. 3991, pp. 611–617. Springer, Heidelberg (2006)

12. Yianilos, P.N.: Data structures and algorithms for nearest neighbor search in general metric spaces. In: Proceedings of the Fourth Annual Symposium on Discrete Algorithms, pp. 311–321 (1993)

13. Zezula, P., Amato, G., Dohnal, V., Batko, M.: Similarity Search: The Metric Space Approach. Springer (2006)

# DSACL+-tree: A Dynamic Data Structure for Similarity Search in Secondary Memory

Luis Britos, A. Marcela Printista, and Nora Reyes

Dpto. de Informática, Universidad Nacional de San Luis,
Ejército de los Andes 950, San Luis, Argentina
{lebritos,mprinti,nreyes}@unsl.edu.ar

**Abstract.** Metric space searching is an emerging technique to address
the problem of efficient similarity searching in many applications, including multimedia databases and other repositories handling complex
objects. Although promising, the metric space approach is still immature
in several aspects that are well established in traditional databases. In
particular, most indexing schemes are not dynamic. From the few dynamic indexes, even fewer work well in secondary memory. That is, most
of them need the index in main memory in order to operate efficiently. In
this paper we introduce two different secondary-memory versions of the
Dynamic Spatial Approximation Tree with Clusters (*DSACL-tree* from
Barroso et al.) which has shown to be competitive in main memory. These
two indexes handle well the secondary memory scenario and are competitive with the state of the art. But in particular the innovations proposed
by the version *DSACL+-tree* lead to significant performance improve-
ments.The resulting data structures can be useful in a wide range of
database application.

**Keywords:** dynamic metric indexes, secondary memory.

## 1  Introduction

As the growth of digital data accelerates in variety and extend, the contemporary databases are bulkier and more complex in nature. To manage this bulk
and complexity increasing new techniques are employed, with the multimedia
data for example, the standard approach is to search not at the level of the
actual multimedia objects, but rather using characteristics extracted from these
objects. In such environments, an exact match has little meaning, a very useful search paradigm is to quantify the proximity, similarity, or dissimilarity of
a query object versus the objects stored in a database to be searched. Similarity or proximity searching have became a fundamental computational tasks
with application in many areas as non-traditional databases, data mining, machine learning, data compression; and so on. A useful abstraction for nearness is
provided by the mathematical notion of *metric space*.

In a metric space, there is a universe $U$ of objects and a nonnegative function
$d: U \times U \to \mathbb{R}^+$ defined among them, that will denote a measure of *"distance"*

G. Navarro and V. Pestov (Eds.): SISAP 2012, LNCS 7404, pp. 116–131, 2012.

between objects. This distance function satisfies the three axioms that make $(U, d)$ a *metric space*: *strict positiveness* $(d(x, y) \geq 0$ and $d(x, y) = 0 \Leftrightarrow x = y)$, *symmetry* $(d(x, y) = d(y, x))$, and *triangle inequality* $((d(x, z) \leq d(x, y) + d(y, z)))$. The smaller the distance between two objects, the more *"similar"* they are. A finite subset $X \subseteq U$ with size $n = \mid X \mid$, is called *database* and represents the collection of objects. We are interested to answer *similarity queries* posed to this database. That is, given a new object from the universe (a query) $q \in U$, we must retrieve all the elements similar enough to the query in the database. The database is preprocessed so as to build an *index* that reduces query time. There are two typical queries of this kind:

**Range Query:** Retrieve all elements within distance $r$ to $q$ in $S$. This is, the set $\{x \in S, d(x, q) \leq r\}$.

**Nearest Neighbor Query ($k$-NN):** Retrieve the $k$ closest elements to $q \in S$. That is, a set $A \subseteq S : |A| = k$ and $\forall\ x \in A, y \in S - A, d(x, q) \leq d(y, q)$.

In this paper we are devoted to range queries. Nearest neighbor queries can be rewritten as range queries in an optimal way [9], so we can restrict our attention to range queries. In order to answer queries efficiently the database is preprocessed so as to build an *index* that reduces query time. This metric space approach to similarity search is becoming widely popular [17, 22] and a large number of indexing methods have flourished [4, 10, 8], but mature solutions from the database viewpoint are a long way off.

Most of the existing indexes are *static*: Once they are built for a given dataset, adding more elements, or removing an element from it, requires an expensive updating of the index [13]. Some indexes tolerate insertions in principle, but their quality degrades and require periodic rebuildings, [19–21] among others. There are some structure parameters that may depend on $n$ and thus require periodical structural reorganization (e.g., adding or removing pivots is generally problematic). Others tolerate deletions with the same quality degradation problem, [3, 6, 2, 20, 21] to name a few. Thus there are few *dynamic* indexes.

There are also many interesting databases for similarity searching where the objects are so large that they must stay on disk; or the objects are so many that the index itself cannot fit in main memory. The total time to evaluate a query, as it is explained in [4], can be split as: $T =$ distance evaluations $*$ complexity of $d()$ + extra CPU time + I/O time, and we would like to minimize $T$. In many applications, however, evaluating $d()$ is so costly that the other components of the cost can be neglected. In this case, although the similarity computation can be expensive (e.g., taking milliseconds of CPU time) we cannot disregard disk costs. Therefore, we use this model in this article, and hence the number of distance evaluations performed jointly with the number of I/O operations are the measure of the complexity of the algorithms.

Algorithms to search in general metric spaces can be divided into two large areas: *pivot-based* algorithms and *compact partition-based* ones. Pivot-based algorithms are better suited for low dimensional metric spaces, while compact partitions ones deal better with high dimensional metric spaces. Although the former can improve by using more memory, they need more and more memory to beat the latter as dimension grows. On the other hand, indices based on compact partitions use a fixed amount of memory and cannot be improved by giving them more space. However, there are algorithms that combine ideas from both areas. For lack of space, we do not cite all the previous works, however complete surveys about existing methods can be found in [17, 22, 4, 10, 8].

From the few existing dynamic indexes, even fewer work well in secondary memory. That is, most of them need the data structure in main memory in order to operate efficiently. Although for some applications a static scheme may be acceptable, many relevant ones do require dynamic capabilities. Actually, in many cases it is sufficient to support insertions, such as in digital libraries and archival systems, versioned and historical databases, and several other scenarios where objects are never updated or deleted. There are some index specifically designed for secondary memory and with dynamic capabilities, such as the famous *M-tree* [5], the *D-index* [7], the *PM-tree* [18, 12], and *EGNAT* [16].

In this paper we introduce a new dynamic index aimed at secondary memory. We base our work on that of Barroso et al, called Dynamic Spatial Approximation Tree with Clusters (*DSACL-tree*) [1]. They have shown that the *DSACL-tree* gives an attractive tradeoff between memory usage, construction time, and search performance. Our secondary memory versions, retains these good features, and in addition perform well in secondary memory. We focus on handling insertions and searches in this paper, leaving deletions for future works.

## 2    Dynamic Spatial Approximation Trees

In this section we will describe briefly the *Dynamic Spatial Approximation Tree* (*DSA-tree*), in particular the version called *timestamp with bounded arity* (reported in [14] as one of the better options for this dynamic tree), on top of which *DSACL-tree* [1] was built. The *DSA-tree* is a data structure to answer similarity queries in metric spaces based on the concept to approach the query spatially, getting closer and closer to it, so when we look for an element from the universe (a query $q \in U$) and being in some element $a$ belonging to the database $S$ ($S \subseteq U$), the goal is to move to another object of $S$ spatially closer of $q$ than $a$. When it is not possible to do this move anymore, we are positioned on the element closest to $q$ from $S$.

The *DSA-tree* is built incrementally via insertions. The tree has a maximum arity $A$. Each tree node $a$ stores a *timestamp* of its insertion time, $time(a)$, its *covering radius*, $R(a)$, and its set of children $N(a)$ (the *neighbors* of $a$). To insert a new element $x$, its point of insertion is sought starting at the tree root and moving to the neighbor closest to $x$, updating $R(a)$ in the way. We finally insert $x$ as a new (leaf) child of $a$ if **(1)** $x$ is closer to $a$ than to any $b \in N(a)$, and **(2)** the arity of $a$, $|N(a)|$, is not already maximal. In other case, we insert $x$ in the subtree of the closest element $b \in N(a)$. Neighbors are stored left to right in increasing timestamp order. Note that each element is older than its children and than its next sibling.

The idea for range searching is to replicate the insertion process of relevant elements. That is, we act as if we wanted to insert $q$ but keep in mind that relevant elements may be at distance up to $r$ from $q$. So in each decision for simulating the insertion of $q$ we permit a tolerance of $\pm r$, so that it may be that relevant elements were inserted in different children of the current node, and backtracking is necessary.

We have to consider two facts, at the time an element $x$ was inserted. The first is that, a node $a$ in its path may not have been chosen as its parent because its arity was already maximal. So, at query time, instead of choosing the closest to $x$ among $\{a\} \cup N(a)$, we may have chosen only among $N(a)$. Hence, we perform the minimization only among elements in $N(a)$. The second fact is that, elements with higher timestamp were not yet present in the tree, so $x$ could choose its closest neighbor only among elements older than itself.

Hence, we consider the neighbors $\{b_1, \ldots, b_k\}$ of $a$ from oldest to newest, disregarding $a$, and perform the minimization as we traverse the list. This means that we enter into the subtree of $b_i$ if $d(q, b_i) \leq \min\{d(q, b_1), \ldots, d(q, b_{i-1})\} + 2r$.

Up to now we do not really need the exact timestamps but just to keep the neighbors sorted by timestamp. We make better use of the timestamp information in order to reduce the work done inside older neighbors. Say that $d(q, b_i) > d(q, b_{i+j}) + 2r$. We have to enter into the subtree of $b_i$ anyway because $b_i$ is older. However, only the elements with timestamp smaller than that of $b_{i+j}$ should be considered when searching inside $b_i$; younger elements have seen $b_{i+j}$ and they cannot be interesting for the search if they are inside $b_i$. As parent nodes are older than their descendants, as soon as we find a node inside the subtree of $b_i$ with timestamp larger than that of $b_{i+j}$ we can stop the search in that branch, because all its subtree is even younger.

## 3   Dynamic Spatial Approximation Trees between Clusters

In this section we will describe briefly the *Dynamic Spatial Approximation Trees between Clusters* (*DSACL-tree*) [1]. The *DSACL-tree* performs the spatial approximation on groups or *clusters* of objects that are very close to each other, rather than individual objects. By this way it can reduce search costs, because it has to do less backtracking. Therefore, in the *DSACL-tree* each node represents a cluster of very similar objects, for short we refer to it simply as *cluster*. Thus, we relate the clusters by their proximity in the metric space. So, each node of the tree would be able to store multiple database objects, reducing the number of nodes with respect to the original *DSA-tree*.

As in the *DSA-tree* we build the tree incrementally, considering a *maximum arity* and maintaining information of the *timestamp* (time of insertion of each element). We also register the *timestamp* $time(c)$ of each node $c$ in the tree, that is the time when this node was created. Each node $c$ keeps an object $center(c)$ as the *center* of the cluster and the $k$ *nearest objects* ($cluster(c)$) seen in its subtree, and is connected with their clusters-neighbors $N(c)$. The cluster also has a *cluster radius* $rc(c)$, that is considering the objects in increasing order to the $center(c)$ the distance of the $k$-th object in the $cluster(c)$. Any object further away from the center than $rc(c)$ would become part of another tree node, which could be a new neighbor in some cases, since the arity is bounded in the same way as *DSA-tree*. Each node $c$ also stores the maximum distance between the *center(c)* and the farthest object in its subtree $R(c)$ (as *DSA-tree* does), called *covering radius* of the subtree of $c$. Figure 1(b) shows some details of an example of *DSACL-tree* in the subspace of $\mathbb{R}^2$, in order to illustrate the concepts of the covering radius $R(c)$ of a node $c$, the cluster radius $rc(c)$, and the cluster of each object *cluster(c)*, the neighbors of a node $c$ are the other nodes pointed from $c$. For this example the maximum arity is 3 and maximum cluster size is 2.

Since each node $c$ represents a cluster centered in *center(c)* with at most $k$ objects within *cluster(c)*, we maintain the distances between *center(c)* and all the objects in *cluster(c)* ordered by increasing distance to the center. At search time, we can use these stored distances in order to minimize the number of distance computations using the triangle inequality. Besides, if $x_1, \ldots, x_k$ are the objects in *cluster(c)* sorted by distances, the covering radius of the cluster will be $rc(c) = d(center(c), x_k)$. Therefore, for each object $x_i$ inside the cluster, we stored its insertion moment $time(x_i)$ and the distance $d(center(c), x_i)$. It is clear that it is not necessary to really register $rc(c)$ because it can be obtained from the stored distances inside the node. During searches, both radios $rc(c)$ and $R(c)$ are used to rule out entire areas of space containing non relevant elements.

Because of the spatial approximation, to insert a new element $x$, we should go down the tree until found the node $c$ such that $x$ is closer to *center(c)* than the centers of neighbors in $N(c)$. If in *cluster(c)* there is room for one more element, then it will be inserted along with its distance. If there is no room, we must choose the most distant element $b$ among the $k$ elements in *cluster(c)* and $x$ ($k + 1$-th in distance order from *center(c)*). We have two possible cases:

1. if $b$ is $x$, then $x$ must be added like center of a new neighbor node of $c$, if the arity allows it, otherwise it must choose the node among all the neighbors in $N(c)$ whose center is the nearest and keep the insertion from there.
2. If $b$ is not $x$, then $b$ must choose the nearest center $a$ among *center(c)* and the center of all nodes neighbors in $N(c)$ that are newer than $b$ because when $b$ was inserted, it wasn't compared with them. Later, if $a$ is *center(c)*, the process followed is the same as when $b$ is $x$; otherwise, if $a$ is not *center(c)*, then continues with the insertion of $b$ from the node with center $a$.

Algorithm 1 illustrates the whole insertion process. The function is invoked as InsertCl($a$, $x$), where $a$ is the root node and $x$ is the element to be inserted.

When performing a range query, we proceed in a similar way as *DSA-tree*, that is we perform the spatial approximation to the query via the centers of nodes. As we mentioned previously, the idea for range searching is to replicate the insertion process of the relevant elements to the query. That is, we act as if we wanted to insert $q$ but keeping in mind that relevant elements may be at distance up to $r$ from $q$, so in each decision we simulate the insertion of $q$ permitting a tolerance of $\pm r$. So that it may be that relevant elements were inserted in a cluster, in different children of the current node, and backtracking is necessary.

**Algorithm 1.** Insertion algorithm of a new element $x$ in a *DSACL-tree* with root node $a$

```
InsertCl (Node a, Element x)
 1.  R(a) ← max(R(a), d(center(a), x))
 2.  If (((|cluster(a)| < k) ∨ (d(center(a), x) < rc(a))) Then
 3.     cluster(a) ← cluster(a) ∪ {x}
 4.     d'(x) ← d(center(a), x)
 5.     timestamp(x) ← CurrentTime
 6.     If (|cluster(a)| = k + 1) Then
 7.        y ← argmax z ∈ cluster(a) d'(z)
 8.        cluster(a) ← cluster(a) − {y}
 9.        InsertCl(a, y)
10.  Else
11.     c ← argmin_{b∈N(a)} d(center(b), x)
12.     If d(center(a), x) < d(center(c), x) ∧ |N(a)| < MaxArity
            Then /* b is a new node, neighbor of a, with center(b) = x */
13.        N(a) ← N(a) ∪ {b}
14.        center(b) ← x
15.        N(b) ← ∅, R(b) ← 0
16.        cluster(b) ← ∅
17.        timestamp(x) ← CurrentTime
18.        time(b) ← CurrentTime
19.     Else
20.        InsertCl (c, x)
```

In principle, at search time for $q$ with radius $r$, one should report the root a if $d(a, q) \leq r$, then find the closest element $b \in \{center(c)\} \cup N(a)$. Yet, because of the timestamped insertion process, we have to consider the neighbors $b_1, hdots, b_k$ of a from oldest to newest. This is because, between the insertion of $b_i$ and $b_{i+j}$, there may have appeared new elements that preferred to be inserted lower index in $b_i$ is missing because $b_{i+j}$ was not yet a neighbor, so we may miss an element if we do not enter $b_i$ because of the existence of $b_{i+j}$. Only the elements with timestamp smaller than that of $b_{i+j}$ should be considered when searching inside $b_i$; younger elements have seen $b_{i+j}$ and they cannot be interesting for the search if they chose $b_i$. As parent nodes are older than their descendants, as soon as we find a node inside the subtree of $b_i$ with timestamp larger than that of $b_{i+j}$ we can stop the search in that branch, because its subtree is even younger.

## 4   Secondary Memory

The distance is assumed to be expensive to compute. However, when we work in secondary memory, the complexity of the search must consider both the number of distance evaluations performed and the number of I/O operations; other components such as CPU time for side computations can usually be disregarded. Given a dataset of $|S| = n$ objects of total size $N$ and disk page size $B$, queries can be trivially answered by performing $n$ distance evaluations and $N/B$ I/Os.

---

**Algorithm 2.** Range query algorithm on a *DSACL-tree* with root node $a$

---

RangeSearchCl (Node $a$, Query $q$, Radius $r$, Timestamp $t$)

1.  If $time(a) < t \wedge d(center(a), q) \leq R(a) + r$ Then
2.     If $d(center(a), q) \leq r$ Then Report $a$
3.     If $(d(center(a), q) - r \leq rc(a)) \vee (d(center(a), q) + r \leq rc(a))$ Then
4.      For $c_i \in cluster(a)$ Do
5.       If $|(d(center(a), q) - d(c_i)| \leq r$ Then
6.        If $d(c_i, q) \leq r$ Then Report $c_i$
7.     If $d(center(a), q) + r < rc(a)$ Then Return
8.     $d_{min} \leftarrow \infty$
9.     For $b_i \in N(a)$ in increasing order of timestamp Do
10.     If $d(center(b_i), q) \leq d_{min} + 2r$ Then
11.      $k \leftarrow \min\{j > i, d(center(b_i), q) > d(center(b_j), q) + 2r\}$
12.      RangeSearchCl $(b_i, q, r, time(b_k))$
13.     $d_{min} \leftarrow \min\{d_{min}, d(center(b_i), q)\}$

---

The goal of an index is to preprocess the dataset so as to answer queries with as few distance evaluations and I/O operations as possible. We show only the number of I/O operations, even if we are aware that a large number of factors affecting the final speed, because we have limited access to high speed architectures, operating systems and modern hardware.

In *DSACL-tree* each node has a fixed size, because of this, the structure seems to be naturally adequate for secondary memory. In this paper, we propose two different implementations of *DSACL-tree* having as goal achieve good performance in secondary memory without compromise the good features of the original structure. As it can be seen, the main difference between both implementations is the way it manages and stores the neighbors set $N(a)$ in the disk.

These two implementations make also a partition of the searching space considering spatial proximity, grouping the closest elements, relating complete clusters by its proximity in the space. This permits that each node of the structure is capable of storing multiple elements from the database. Also, to avoid disk underutilization, we will fix the size of clusters and also the maximum arity in function to the page size available, therefore, each node takes exactly one page in disk, simplifying the administration of nodes.

### 4.1   DSACL*-tree and DSACL+-tree

The first version which we call the *DSACL*-tree* is a direct implementation of *DSACL-tree*, this means that the nodes will have exactly the same estructure of its version in primary memory (to see Figures 1(a), 1(b), and 1(c)). The *DSACL*-tree* implements $N(a)$ as a linked list (using the binary tree representation [11]), thus, each node contains two pointers, the first pointer to the first neighbors and the second pointer to the rest of the neighbors of the set.

To insert an element $x$ into the *DSACL*-tree* structure we find the insertion point in the tree, following a unique path, so that when we determine that $x$ should be added to a node $a$ because $x$ is closer to $a$ than to any neighbor in

$N(a)$, already have loaded the page corresponding to $a$. If $a$ already have its $k$ elements then it must choose the element furthest from the center from its $k + 1$ elements and then choose if $x$ must to be inserted like center of a new neighbor, if the arity allows it, otherwise the insertion must to continue forcing $x$ to choose the closest neighbor from $N(a)$ and keeping going down on the tree recursively. This implementation saves space, since booking room only for two pointers can completely recover the set $N(a)$ making it possible to accommodate more objects in each cluster.

The second version which we call *DSACL+-tree* introduces changes to the structure of the node with the aim of decrease the number of I/O in the processes of construction and search (to see Figure 1(d)).

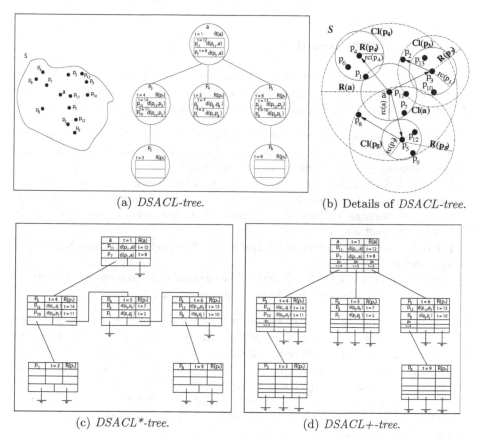

(a) *DSACL-tree.*    (b) Details of *DSACL-tree.*

(c) *DSACL\*-tree.*    (d) *DSACL+-tree.*

**Fig. 1.** Example of a *DSACL-tree* and corresponding *DSACL\*-tree* and *DSACL+-tree*

In *DSACL\*-tree*, when the algorithm needs to access to the node $a$ and its neighbors $N(a)$, it takes $|N(a) + 1|$ inputs. To avoid some I/O operations when the element to insert must to decide which element is closer the center of the node or some neighbor, in the *DSACL+-tree* for each neighbor of a node $a$, we will save its object center, its location on the file and its insertion time.

Note that during an insertion operation, in both versions, the involved pages are accessed with a disk read operation. The pages that need to upgrade a coverage radius by inserting the new object, must also to perform a disk write operation. However, during the search process, the pages are accessible only to be read, since the search can not change anything in the tree. Searches in both proposals are performed as in *DSACL-tree*, which was shown in Algorithm 2.

Because we need to set the size of a node as a size of a page disk, considering the size needed to represent an element we must to fix the maximum arity and the cluster size of each node. If the elements are big it is possible to notice that the arity and the cluster size will be small. However, as it has been demonstrated in [14, 15], it is not a drawback because small arities were a key factor, for the *DSA-tree*, to reduce construction and search costs in several metric spaces.

## 5    Experimental Results

In order to give a broad picture of the performance of our indexes, we had selected four widely different metric spaces, all from the SISAP Metric Space Library (www.sisap.org). The metric spaces considered were:

- WORDS: a dictionary of 69,069 English words. The distance is the *edit distance*, that is, the minimum number of character insertions, deletions and substitutions needed to make two strings equal.
- DOCUMENTS: 1,265 documents under the Cosine similarity, from TREC-3 collection. In this model the space has one coordinate per term and documents are seen as vectors in this space. The distance we use is the angle among the vectors.
- IMAGES: 40,700 20-dimensional feature vectors, generated from NASA images, using Euclidean distance.
- HISTOGRAMS: 112,682 8-D color histograms(112 - dimensional vectors) from an image database. Euclidean distance is used.

For search experiments, we built the indexes with 90% of the objects and used the other 10% (randomly chosen) as queries. All our results are averaged over 10 index constructions using different database permutations. We have considered range queries retrieving on average 0.01%, 0.1% and 1% of the dataset. This corresponds to radii 0.14, 0.15 and 0.195 for DOCUMENTS, 0.12, 0.285 and 0.53 for IMAGES, and 0.051768, 0.082514 and 0.131163 for HISTOGRAMS. WORDS have a discrete distance, so we used radii 1 to 4. The same queries were used for all the experiments on the same datasets.

In order to compare fairly the two versions of *DSACL-tree*, for each metric space considered, we need to select the best maximum arity. However, for lack of space we do not show here the comparison of behaviors for the different arities used, for search and construction costs. As we mention previously, the maximum arity allowed also affects the cluster size, that is there exists a tradeoff between maximum arity and cluster size and this tradeoff affects the number of I/O operations performed: *if the arity was small, the cluster can increase its size, and it has showed to be good to minimize the I/O operations.* Moreover, for the same arity the cluster size of *DSACL+-tree* can be smaller than that of *DSACL\*-tree*, because *DSACL+-tree* also stores the centers of its neighbors.

The construction and query costs are compared regarding the number of distance evaluations and the number of I/O operations because they are the most expensive operations involved, and they are representative of real costs. Besides, we can infer the approximate costs regardless of the computer used.

The best arity is determined mainly by its search performance considering both the number of distance calculations and the number of I/O operations required, but if the search performances were similar we select the arity that obtains better construction cost. The best arities obtained for *DSACL\*-tree*, for each metric space considered, are: arity 2 in WORDS, 8 in DOCUMENTS, 4 in IMAGES and 2 in HISTOGRAMS. The best arity obtained for *DSACL+-tree*, for each metric space considered, are: arity 4 in WORDS, 4 in DOCUMENTS, 4 in IMAGES and 2 in HISTOGRAMS.

## Comparing vs other Structures

In this section we compare our structures with some structures representatives of state of art, they are *DSA\*-tree* and *DSA+-tree* [15], and *M-tree* [5], whose codes are available. For a fair comparison, we have chosen the parameters that

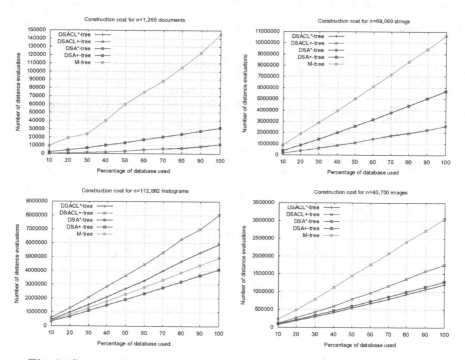

**Fig. 2.** Construction costs, for all indexes, comparing distance evaluations

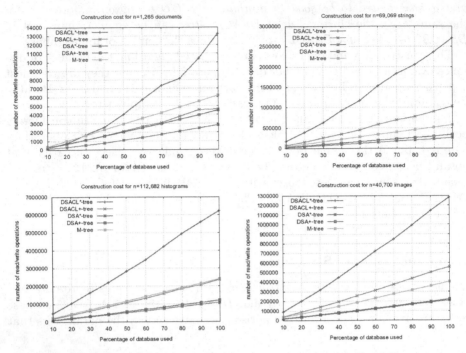

**Fig. 3.** Construction costs, for all indexes, comparing the number of I/O operations

**Table 1.** Fill ratio for the different datasets

| | Fill ratio | | | |
| Dataset | DSA*-tree | DSA+-tree | DSACL*-tree | DSACL+-tree |
|---|---|---|---|---|
| WORDS | 83% | 66% | 69% | 70% |
| DOCUMENTS | 84% | 68% | 68% | 69% |
| IMAGES | 80% | 67% | 72% | 73% |
| HISTOGRAMS | 75% | 67% | 66% | 91% |

**Table 2.** Total pages used for the different datasets

| | Total pages used | | | | |
| Dataset | DSA*-tree | DSA+-tree | M-tree | DSACL*-tree | DSACL+-tree |
|---|---|---|---|---|---|
| WORDS | 904 | 1,536 | 1,608 | 885 | 901 |
| DOCUMENTS | 12 | 22 | 31 | 9 | 9 |
| IMAGES | 1,271 | 1,726 | 1,973 | 1,310 | 1,366 |
| HISTOGRAMS | 18,781 | 21,136 | 31,791 | 18,827 | 24,827 |

showed the best performance of the structures with which we compare against, and experiments have exactly the same characteristics as those explained in the previous section for experimentation performed on the *DSACL*-tree* and

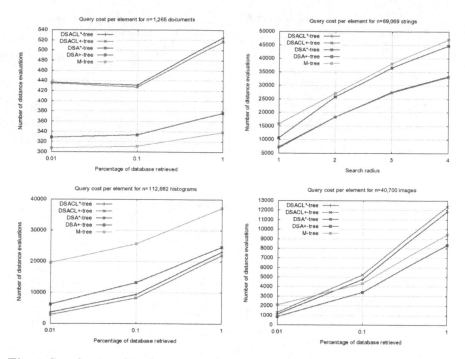

**Fig. 4.** Search costs, for all indexes, comparing the number of distance evaluations

*DSACL+-tree.* We built the *DSA\*-tree* and *DSA+-tree* by successive insertions using a 4-arity for all spaces except for the WORDS space where we use a 32-arity. Searches were used for the same radii as those used for *DSACL\*-tree* and *DSACL+-tree*. The *M-tree* construction is performed on the same way, through successive insertions, and used the following values for the parameters: Split Function: Generalized Hyperplane, part Promote Function: MIN RAD, Secondary part Function: MIN RAD Radius Function: LB and Min Util: 0.2. In all cases we use the same page size of 4KB.

Figure 2 and Figure 3 show the comparison between all the alternative indexes considering construction costs, for all metric spaces, but Figure 2 comparing distance computations and Figure 3 number of I/O operations. It can be noticed that our proposals outperform the others indexes regarding distance evaluations, in most of the metric space considered. However, despite of our data structures have the higher number of read/write operations, they have to read/write in more compact files. Therefore, this behavior do not affect so much the construction performance.

Figure 4 and Figure 5 depict the comparison between search costs of all the alternative indexes and for all metric spaces, but Figure 4 compares distance computations and Figure 5 the number of page read, because at search time there is no write operation. As regarding the number of distance evaluations our indexes outperform the others one in only two of the metric spaces considered, but the best option is our *DSACL+-tree* in the case of the number of read/write

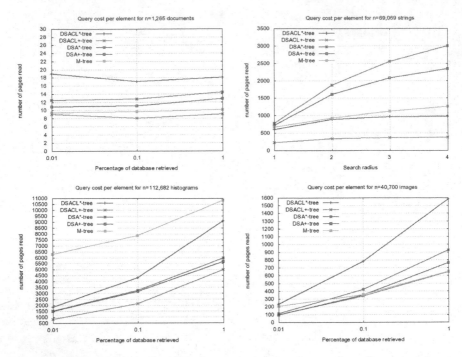

**Fig. 5.** Search costs, for all indexes, comparing the number of pages read

operations because it beats all indexes. Therefore, our *DSACL+-tree* can be a very good index compared against the other competitive indexes.

Other important aspect to analyze is that the file that holds the index was as compact as possible; i.e. having a high occupancy rate (not having too empty pages). As it can be noticed that our structures have a good fill ratio, in all cases over 65%. Table 1 and Table 2 show the average disk page occupancy achieved for all metric spaces. We compare the fill ratio and the total number of pages used by *DSA\*-tree*, *DSA+-tree* and *M-tree* against our results. Our structure *DSACL+-tree* obtains a good fill ratio and uses, in general, fewer disk pages than some of the other indexes designed for secondary memory, while maintains a good search performance.

Following with the experimental evaluation of our proposals, we consider two bigger metric spaces:

- VOCABS: the subset of 494,048 terms of a documents collection belonging to Wall Street Journal. The distance is the *edit distance*.
- GAUSS: a synthetic metric space that contains 1,000,000 vectors of 50 real coordinates, created by using a Gaussian distribution with different mean values and a variance equal to 0.1, and grouped in 10,000 clusters. We use in this case Euclidean Distance. This space was generated with the codes available at SISAP, from the Development Guide of the Sixth Annual DIMACS Implementation Challenge: Near Neighbor Searches.

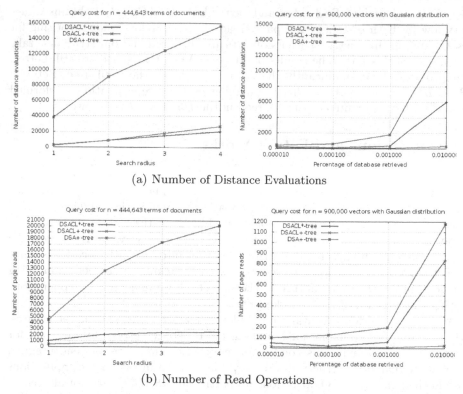

(a) Number of Distance Evaluations

(b) Number of Read Operations

**Fig. 6.** Search costs, for *DSACL\*-tree*, *DSACL+-tree*, and *DSA+-tree*

We only show the search experiments. As before, we built the indexes with 90% of the objects and used the other 10% (randomly chosen) as queries and our results are averaged over 10 index constructions using different database permutations. We have considered range queries retrieving on average 0.00001%, 0.0001%, 0.001%, and 0.01% of the dataset. This corresponds to radii for GAUSS of 2.77, 3.7, 6.9, and 56. VOCABS have a discrete distance, so we used radii 1 to 4. The same queries were used for all the experiments on the same datasets.

In the Figure 6 we illustrate the searches behavior only for *DSACL\*-tree*, *DSACL+-tree*, and *DSA+-tree*. Figure 6(a) depicts the number of distance evaluations and Figure 6(b) the number of read operations, performed for the three methods. We select for these indexes the best value of their parameters. For GAUSS metric space *DSACL\*-tree* and *DSACL+-tree* both use arity 2, and *DSA+-tree* uses arity 4, while for VOCABS *DSACL\*-tree*, *DSACL+-tree*, and *DSA+-tree* uses arity 16, arity 2, and arity 32 respectively. We do not evaluate the other indexes here for lack of time.

For GAUSS space, it can be observed that for small query radii all structures shows similar results, but as radii grows it can be better appreciated that our structures outperform *DSA+-tree* as much in distance evaluations and number of pages read. Clearly our both structures show a better behavior when they are compared against *DSA+-tree* and particularly *DSACL+-tree* obtains values very low for both distances and pages read. This behavior is not surprising for us, because in this metric space its objects are clustered and our structures take advantage of this fact. The results obtained in VOCABS space are still better, our proposals beat significantly the *DSA+-tree*, in all the radii used. Between *DSACL\*-tree* and *DSACL+-tree* the number of distance evaluations performed is similar, but the number of pages read in the *DSACL+-tree* is lower.

## 6   Conclusions and Future Works

In this work we present the *DSACL\*-tree* and the *DSACL+-tree* both indexes for searching metric spaces for secondary memory. These new indexes enhance the good features of the *DSACL-tree* (spatial approximation, dynamism, and clustering), and also take into account the I/O operations costs. In fact, each node of our structures corresponds to a page. By this way, we try to get the most advantage in each read or write operation into the disk, locating similar objects together. Therefore, we reduce the backtracking at searches improving the cost, in distance evaluations, at the same time we make few I/O operations during the retrieval of relevant elements.

We compared experimentally our indexes against other existing structures, whose codes are available, for searching metric spaces especially designed for achieve a good behavior in secondary memory. The results are very encouraging, they show that both structures are very competitive in distance computations and in the number of I/O operations. In particular *DSACL+-tree*, for all spaces, gets the least amount of I/O operations, in addition, gets the least number of distance evaluations in three of the six spaces considered (WORDS, HISTOGRAM and GAUSS). Then, the *DSACL+-tree* stands out as a practical and efficient data structure that can be used in a wide range of applications,while retaining the good features of the original data structure, specially thinking in secondary memory. The most important remaining work is to handle deletions. Besides, another interesting problem to address is the design of bulk-loading mechanisms.

Moreover, for future works, we plan to deeply analyze the scalability of these methods, by complementing our experimentation with new massive metric spaces in addition to considering the variation of other parameters sensitive to the problem, as is the size of the disk page.

## References

1. Barroso, M., Reyes, N., Paredes, R.: Enlarging nodes to improve dynamic spatial approximation trees. In: Proceedings of the 3rd International Conference on Similarity Search and Applications (SISAP 2010), pp. 41–48. ACM Press (2010), doi: http://doi.acm.org/10.1145/1862344.1862351
2. Brin, S.: Near neighbor search in large metric spaces. In: Proc. 21st Conference on Very Large Databases (VLDB 1995), pp. 574–584 (1995)
3. Burkhard, W., Keller, R.: Some approaches to best-match file searching. Comm. of the ACM 16(4), 230–236 (1973)

4. Chávez, E., Navarro, G., Baeza-Yates, R., Marroquín, J.: Searching in metric spaces. ACM Comput. Surv. 33(3), 273–321 (2001)
5. Ciaccia, P., Patella, M., Zezula, P.: M-tree: an efficient access method for similarity search in metric spaces. In: Proc. 23rd Conf. on Very Large Databases (VLDB 1997), pp. 426–435 (1997)
6. Dehne, F., Noltemeier, H.: Voronoi trees and clustering problems. Information Systems 12(2), 171–175 (1987)
7. Dohnal, V.: An Access Structure for Similarity Search in Metric Spaces. In: Lindner, W., Fischer, F., Türker, C., Tzitzikas, Y., Vakali, A.I. (eds.) EDBT 2004. LNCS, vol. 3268, pp. 133–143. Springer, Heidelberg (2004)
8. Hetland, M.L.: The Basic Principles of Metric Indexing. In: Coello, C.A.C., Dehuri, S., Ghosh, S. (eds.) Swarm Intelligence for Multi-objective Problems in Data Mining. SCI, vol. 242, pp. 199–232. Springer, Heidelberg (2009)
9. Hjaltason, G., Samet, H.: Incremental similarity search in multimedia databases. Tech. Rep. CS-TR-4199, University of Maryland, Computer Science Dept. (2000)
10. Hjaltason, G., Samet, H.: Index-driven similarity search in metric spaces. ACM Trans. on Database Systems 28(4), 517–580 (2003)
11. Knuth, D.E.: The Art of Computer Programming, Volume I: Fundamental Algorithms, 2nd edn. Addison-Wesley (1973)
12. Lokoc, J., Skopal, T.: On reinsertions in m-tree. In: SISAP 2008: Proceedings of the First International Workshop on Similarity Search and Applications (SISAP 2008), pp. 121–128. IEEE Computer Society, Washington, DC (2008)
13. Navarro, G.: Searching in metric spaces by spatial approximation. The Very Large Databases Journal 11(1), 28–46 (2002)
14. Navarro, G., Reyes, N.: Dynamic spatial approximation trees. ACM Journal of Experimental Algorithmics 12, article 1.5, 68 pages (2008)
15. Navarro, G., Reyes, N.: Dynamic spatial approximation trees for massive data. In: Proc. 2nd International Workshop on Similarity Search and Applications (SISAP), pp. 81–88. IEEE CS Press (2009)
16. Navarro, G., Uribe, R.: Fully dynamic metric access methods based on hyperplane partitioning. Information Systems 36(4), 734–747 (2011)
17. Samet, H.: Foundations of Multidimensional and Metric Data Structures (The Morgan Kaufmann Series in Computer Graphics and Geometric Modeling). Morgan Kaufmann Publishers Inc., San Francisco (2005)
18. Skopal, T., Pokorný, J., Snásel, V.: PM-tree: Pivoting metric tree for similarity search in multimedia databases. In: ADBIS (Local Proceedings) (2004)
19. Uhlmann, J.: Satisfying general proximity/similarity queries with metric trees. Information Processing Letters 40, 175–179 (1991)
20. Yianilos, P.: Data structures and algorithms for nearest neighbor search in general metric spaces. In: Proc. 4th ACM-SIAM Symposium on Discrete Algorithms (SODA 1993), pp. 311–321 (1993)
21. Yianilos, P.: Excluded middle vantage point forests for nearest neighbor search. In: DIMACS Implementation Challenge, ALENEX 1999, Baltimore, MD (1999)
22. Zezula, P., Amato, G., Dohnal, V., Batko, M.: Similarity Search: The Metric Space Approach. Advances in Database Systems, vol. 32. Springer (2006)

# Scalable Distributed Algorithm for Approximate Nearest Neighbor Search Problem in High Dimensional General Metric Spaces

Yury Malkov, Alexander Ponomarenko, Andrey Logvinov, and Vladimir Krylov

MERA Labs LLC, Nizhny Novgorod, Russia
{ymalkov,aponom,alogvinov,vkrylov}@meralabs.com

**Abstract.** We propose a novel approach for solving the approximate nearest neighbor search problem in arbitrary metric spaces. The distinctive feature of our approach is that we can incrementally build a non-hierarchical distributed structure for given metric space data with a logarithmic complexity scaling on the size of the structure and adjustable accuracy probabilistic nearest neighbor queries. The structure is based on a small world graph with vertices corresponding to the stored elements, edges for links between them and the greedy algorithm as base algorithm for searching. Both search and addition algorithms require only local information from the structure. The performed simulation for data in the Euclidian space shows that the structure built using the proposed algorithm has navigable small world properties with logarithmic search complexity at fixed accuracy and has weak (power law) scalability with the dimensionality of the stored data.

**Keywords:** Similarity Search, Nearest Neighbor, Approximate Nearest Neighbor, Small World, Distributed Data Structure, Metric space.

## 1 Introduction

The scalability of any software system is limited by the scalability of its data structures. Massively distributed systems like BitTorrent or Skype are based on the distributed hash tables. While the latter have good scalability, their search functionality is limited to the exact element hash value matching. This limitation arises because small changes in an element value lead to large and chaotic changes in the hash value, making the hash-based approach inapplicable to the range search and the similarity search problems.

However, there are many applications (such as pattern recognition and classification [1], content-based image retrieval [2], machine learning [3], recommendation systems [4], searching similar DNA sequence [5], semantic document retrieval [6]) that require the similarity search rather than just exact matching. The nearest neighbor search (NNS) problem is a mathematical formalization for the similarity search. It is defined as follows: we need to find the closest object $p \in X$ from a finite set of objects $X \subseteq \mathcal{D}$ to a given query $q \in \mathcal{D}$, where $\mathcal{D}$ is a set of all possible objects (the

G. Navarro and V. Pestov (Eds.): SISAP 2012, LNCS 7404, pp. 132–147, 2012.
© Springer-Verlag Berlin Heidelberg 2012

data domain). Closeness or proximity of two objects $o',o'' \in \mathcal{D}$ is defined as a distance function $\sigma(o',o'')$.

A naïve solution for the NNS problem is to calculate the distance function $\sigma$ between q and every element from X. This leads to linear search time complexity scalability with the number of elements which is much worse than the scalability of structures with the exact value search and makes it almost impossible to use the NNS for extreme size datasets.

We suggest a solution for the nearest neighbor search problem, a data structure with a small world network topology represented by a graph $G(V,E)$, where every object $o_i$ from $X$ is uniquely associated with a vertex $v_i$ from $V$. Searching for the closest element to the query $q$ from the data set $X$ takes a form of searching for a vertex in the graph $G$.

We chose this approach based on the following:

- There are many existing well-developed algorithms for building small world networks for some special cases [7].
- Small world networks principally have no root element.
- All operations (addition and search) use only local information and can be initiated from any element that was previously added to the structure.

This gives an opportunity for building decentralized similarity search oriented storage systems where physical data location doesn't depend on the content because every data object can be placed on an arbitrary physical machine and can be connected with others by links like in p2p systems. Such storage systems can provide a simultaneous access to large numbers of users performing data search and addition, have good fault tolerance and are highly scalable in terms of performance and capacity.

One of the basic vertex search algorithms in graphs with metric objects is the greedy search algorithm. It has a simple implementation and can be initiated from any vertex. In order for a result of the algorithm to be always the exact nearest neighbor to any query, the network must contain the Delaunay graph as its subgraph, which is dual to the Voronoi tessellation [8]. However, there are major drawbacks associated with the Delaunay graph, it requires some knowledge of metric space internal structures [9] and it suffers from the curse of dimensionality [8]. Moreover the requirement of the search for the exact nearest neighbor can be not necessary (optional) for the applications described above. So the problem of finding the exact nearest neighbor can be substituted by the approximate nearest neighbor search, and thus we don't need to support the whole/exact Delaunay graph.

For the greedy search algorithm to be logarithmically scalable, the small world network should have the navigation property [7].

In this paper we present a very simple algorithm for the data structure construction based on a small world network topology with a graph $G(V,E)$ which uses the greedy search algorithm for the approximate nearest neighbor search problem. The graph $G(V,E)$ contains an approximation of the Delaunay graph and has long-range links together with the small-world navigation property. The search algorithm has an ability to adjust the accuracy of search without modification of the structure. Presented algorithms do not use the coordinate representation and do not presume the properties of linear spaces, because they are based only on the metric computation between the objects, and therefore are applicable to data from general metric spaces. It is shown experimentally that the dimensionality dependence is polynomial for a vector data.

## 2    Related Works

All papers that are dedicated to the nearest neighbor search problem can be divided into four categories: centralized nearest neighbor search structures;  centralized approximate exact nearest neighbor search structures, distributed exact nearest neighbor search structures and distributed approximate nearest neighbor search structures.

### 2.1    Centralized Exact Nearest Neighbor Search Structures

Kd-tree[10] and quadra trees[11] were among the first works on the NNS problem. They perform well in 2-3 dimensions (search complexity is close to $O(\log n)$), but the analysis of the worst case for that structures[12] indicates $O(d * N^{1-1/d})$ search complexity, where $d$ is the dimensionality.

Other structures which have a tree topology such as variants of kd-trees, R-trees and structures based on space-filling curves are surveyed in [13]. They also have good performance when searching in a low-dimension ($d < 4$) metric space, but they quickly lose their effectiveness with the increasing number of dimensions [14].

In general, presently there are no methods for effective exact NNS in high-dimensionality metric space. The reason behind this lies in the "curse" of dimensionality [15]. To avoid the curse of dimensionality while retaining the logarithmic scaling on the number of elements, it was proposed to reduce the requirements for the NNS problem solution, making it approximate (ANN).

### 2.2    Centralized Approximate Nearest Neighbor Search Structures

Thus a large number of papers appeared which proposed to search for the nearest neighbor with predefined accuracy ε (ε-NNS). For example, Arya and Mount proposed methods with search complexity $O(\log^3 n)$, but preprocessing requires $O(n^2)$ and the algorithm was applicable only to data from $E^d$ [16] .

Kleinberg proposed two methods [17] for solving ε-NNS. First method requires $O(n \log d)^{2d}$ preprocessing time and query time polynomial in $d, \varepsilon$ , and $\log n$ . The other method with preprocessing time polynomial in $d$, $\varepsilon$ , and $n$, but with query time $O(n + d \log^3 n)$. Also both methods are applicable only to data from $E^d$.

The first algorithms with search complexity polynomial in $d$, $\log n$, $\varepsilon^{-1}$ and polynomial preprocessing time for fixed $\varepsilon$ were proposed  by Indyk and Motwani in [18] and Kushilevitz, Ostrovsky and Rabani in [19]. Indyk and Motwani were the first ones to relax $\varepsilon$-ANN problem to approximate point location in equal balls ($\varepsilon$-PLEB). For the formulation of the problem in $\varepsilon$-PLEB points in metric space expand to the balls with center at this point and radius $(1+ \varepsilon)r$, it is necessary to determine which ball belongs to the query $q$. Also in [18] proposed a second method, which uses the concept of locality-sensitive hashing in regard to formulation of the problem $\varepsilon$-PLEB, with search time $O(n^{1/(1+\varepsilon)})$. This method however requires near quadratic memory (for small $\varepsilon$). In addition, the first method is applicable only for $E^d$, and the second for the Hamming space.

In general, the concept of locality-sensitive hashing has become popular in the last decade to solve the ANN problem. Other works using the concept of locality-sensitive hashing are [20], [21]. But they all have the same major drawback: each algorithm is focused on a narrow class of metrics such as Hamming distance, Jakarta or $l_s$ norms for Euclidean space.

In [22,23] there were proposed non-distributed algorithms for the approximate k-NN problem suitable in general spaces performing well even in case of high dimensionality. The drawback for the ordering permutations index [23]  is that it has a part of search algorithm with a CPU time linear dataset size scaling, and [22] is an essentially static index.

### 2.3    Distributed Exact Nearest Neighbor Search Structures

There are a number of distributed structures that doesn't support nearest neighbor search in general metric spaces but provide search for interval queries in attribute-based (vector) data or simple Euclidian space. MAAN [24], SCRAP[25] , Mercury [26] support multi-dimensional range queries and Voronet [27] is p2p network oriented to search nearest neighbor in $E^2$ based on Voronoi tessellation [8]. Every peer has coordinates in $E^2$ and has links to all neighbors of its Voronoi region. For the logarithmic navigation Voronet supports long-range links.

The only metric-based distributed structures are M-Chord [28], GHT [29] and MCAN[25]. MCAN uses a pivot-based technique to map the high dimensional metric data to an N-dimensional vector space, and then uses CAN protocol as its underlying structured P2P system, however they all suffer from the curse of dimensionality.

### 2.4    Distributed Approximate Nearest Neighbor Search Structures

Authors in [30] explain how to use locality-sensitive hashing scheme for building the structure in a distributed environment. They suggest using a two-level mapping from a d-dimensional space to the peer identifier space. However the lack of versatility inherent to all LSH schemes remains as its main drawback.

Kleinberg's work [7] has shown the possibility of using navigable small world networks for finding the nearest neighbor with the greedy search algorithm. The algorithm relied on long-range links following power-law length distribution for navigation and 2-dimensional lattice for correctness of the results. In Voronet[27] the approach was extended to arbitrary 2-dimensional data by building a two dimensional Delaunay tessellation instead of a regular lattice. In their next work [31] they have weakened the requirements on the exactness of the search in order to avoid the curse of dimensionality for the d-dimension Euclidian space. The algorithm approximates the Delaunay graph by selecting $2d + 1$ neighbors that minimize the volume of the corresponding Voronoi cell. The algorithm is rather complicated; it relies heavily on the quality of the Delaunay graph approximation, it has to be repeated iteratively to reach acceptable accuracy and in principle works only in the Euclidian space. The work also presented some sophisticated algorithms for managing the long range links.

# 3     Structure Definition

The structure $S$ is constructed as a small world network represented by a graph $G(V, E)$, where objects from the set $X$ are uniquely mapped to vertices from the set $V$. The set of edges $E$ is determined by the structure construction algorithm. Since each vertex is uniquely mapped to an element from the set $X$, we will use the terms "vertex", "element" and "object" interchangeably. We will use the term "friends" for vertices that share an edge. The list of vertices that share a common edge with the vertex $v_i$ is called the friend list of the vertex $v_i$.

We use a variant of the greedy search algorithm as a base algorithm for the NNS. It traverses the graph from an element to an element each time selecting the friend closest to the query until it reaches a local minimum. See a detailed description of the algorithm in the section 4.

Links (edges) in the graph serve two distinct purposes. There is a subset of short-range links, which are used as an approximation of the Delaunay graph[8] required by the Greedy Search algorithm. Another subset is the long-range links, which are used for logarithmic scaling of the Greedy Search, they are responsible for the navigation small world properties of the constructed graph similar to the ones in Kleinberg's [7] work. The structure is illustrated on the Fig. 1.

In our work we focus on the approximation of the Delaunay graph and ways to nullify the errors rising from of the approximation. It can be studied independently because there is a very simple and strict way to create long range links for a predefined data set (see the section 5).

All queries in the structure are independent, they can be done in parallel and if the elements are placed randomly on physical computer nodes the processing query load is shared evenly across physical nodes. And the performance of the system (parallel queries per second) is limited only by the number of the nodes.

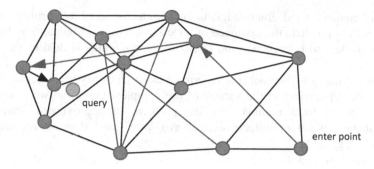

**Fig. 1.** Graph representation of the structure. Circles (vertices) are the data in metric space, **black** edges are the approximation of the Delaunay graph, and **red** edges are long range links for logarithmic scaling. Arrows show a sample path of the greedy algorithm from an enter point to a query (shown green).

# 4    Search Algorithm

## 4.1    Greedy Search

The basic search algorithm traverses the edges of the graph $G(V, E)$ from one vertex to another. The algorithm takes two parameters: query and the vertex $V_{enter\_point} \in V[G]$ which is the starting point of a search (the entry point). Starting from the entry point at each vertex the algorithm computes a metric value from the query q to each vertex from the friend list of the current vertex and then selects a vertex with the minimal metric value. If the metric value between the query and the selected vertex is smaller than the one between the query and the current element, then the algorithm moves to that (new) vertex. The algorithm stops when it reaches a local minimum, a vertex whose friend list doesn't contain a vertex that is closer to the query than the vertex itself. The algorithm:

```
Greedy_Search(q: object, V_enter_point: object)
1   V_curr ← V_enter_point;
2   σ_min ← σ(q, V_curr); V_next ← NIL;
3   foreach V_friend ∈ V_curr.getFriends() do
4       if σ_fr ← σ(query, V_friend) < σ_min then
5           σ_min ← σ_fr;
6           V_next ← V_friend;
7   if V_next = Nil then return V_curr;
8   else return Greedy_Search(q, V_next);
```

The element which is a local minimum with respect to the query $q \in \mathcal{D}$ can be either the true closest element to the query $q$ from the entire set of elements of $X$, or a false closest.

If every element in the structure had in their friend list all of its Voronoi neighbors, then this would preclude the existence of false local minima. Maintaining this condition is equivalent to constructing the Delaunay graph, which is dual to the Voronoi diagram.

It turns out that it is impossible to determine exact Delaunay graph for an unknown metric space [9] (excluding the variant of the complete graph) so we cannot avoid the existence of local minima. For the problem of approximate searching as defined above it is not an obstacle since approximate search does not require the entire Delaunay graph [31].

Note that there is a distinction from the ANN problem defined in the works [16], [17] where it is expressed in terms of $\varepsilon$-neighborhood for which if there are several elements within the $\varepsilon$ of the true nearest neighbor the result of the query can be any of these elements with comparable probabilities. There are no constrains on an absolute value of the distance between the false NN result and true NN result in our structure. Inaccuracy of the algorithm is «topological» in our case, meaning that the most likely result (e.g. with probability 0.95) is the true nearest neighbor, if not, the most likely it will be the second closest and so on with sharply decreasing probability. It may be more convenient to use such definition when the data distribution is highly skewed and it is hard to define one $\varepsilon$ for all regions at the same time.

### 4.2    Multi-search

In order to diminish search errors arising in a network with local minima, we propose a following modification of the search algorithm. We use a series of $m$ searches initiated from random vertices and choose a result element that is closest to the query from the set of found elements. Since the greedy search Greedy_Search(q, $v_{enterPoint}$ $\in$ V) is unambiguous, for each entry point $v_{enterPoint}$ $\in$ V it either results in a success, finding the true nearest neighbor, or in a failure, finding an element that is not the nearest neighbor of $q$.

Thus a search of the closest element to a fixed query $q$ may result in finding the true nearest neighbor (a global minimum) or a false nearest neighbor  depending on the entry point from which the search algorithm started (see Fig. 2).

Since we can choose an entry point at random, there is a probability $p$ of finding the true closest element to a particular element q. Moreover, this probability is always nonzero, because it is always possible to choose the exact nearest neighbor as an entry point, which subsequently will be returned by the greedy search algorithm. As an example the probability of finding query element in Fig. 2 is about 73% since there are 8 elements for which taken as the entry point the algorithm will succeed and 3 elements for which he will not (3/8 results in 73%).

If for a fixed query element probability of finding the true closest in a single search attempt is $p$ then probability of finding the true closest element in at least one of m attempts is $1-(1-p)^m$, thus failure probability decreases exponentially with the number of search attempts. Thus we can improve the search precision, increasing the parameter m - the number of searches from random entry points. For example in the Fig. 2 for m =5 the result probability is 99.985%, which is more than sufficient for the most applications.

The modified greedy search algorithm:

```
Multi_Search(object q, integer: m)
1   results: SET[objects];
2   for (i ← 0; i < m; i++) do
3       entry_point ← getRandomEntryPoint();
4       local_min ← Greedy_Search(query, entry_point)
5       if local_min ∉ results then
6           results.add(result);
7   return results;
```

By selecting the closest element from the results we get an answer to the query.

If m is comparable to the number of elements in the structure, the algorithm becomes an exhaustive search, assuming that entry points are never reused. If the graph of the network has the small-world properties, then it is possible to choose a random vertex in a number of random steps proportional to $\log n$, which doesn't affect the overall logarithmic search complexity. Therefore the overall complexity of a search will increase in no more than m times.

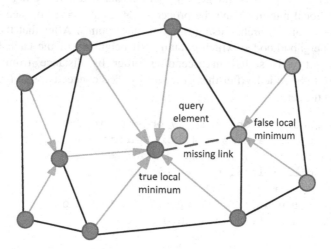

**Fig. 2.** An illustration of the multisearch approach. **Blue** circles represent metric space elements for which taken as entry points for the greedy algorithm it will succeed finding the true NN for a query (**green** circle). **Red** circles represent elements for which taken as entry points the algorithm will stuck in a local minimum. **Arrows** represent gradients direction of the greedy search algorithm. The probability of finding the query in a single search is about 73%. For the multisearch algorithm with m =5 it is 99.985%.

# 5     Data Addition Algorithm

Since we build an approximation of the Delaunay graph, there is a great freedom of choice of the construction algorithm. The main goal of all the works is to minimize the probability of the false local minima while the keeping number of links small. Some approaches are based on knowledge of topology of a metric space being used. For example in [31]  it is proposed to build an approximate Delaunay graph which would minimize a volume of a Voronoi region (computed by the Monte-Carlo method) for a fixed number of edges for each vertex in the graph, this was done by iterating a selection of neighbors of every node in the graph several times. We propose to assemble the structure by adding elements one by one and connecting them on each step with the k closest objects which are already in the structure. It is based on the idea that intersection of the set of elements which are Voronoi neighbors and the k closest elements should be large. Another advantage of this approach was shown empirically in for one-dimensional data[32]. A graph created by such algorithm with data arriving in random order has small world navigation properties without any additional algorithms. That allows us to fully concentrate on the short-range links which approximate the Delaunay graph.

In this work we use a variant of the algorithm which is distinguished by the fact that the search for the k nearest elements uses a series of searches (an analogy to the multi-search, see 4.2).The algorithm takes three parameters: an object to be added to the structure and two positive integer numbers k and w. First, the algorithm determines a set of local minima using the procedure Multi_Search (see 4.2), which produces a series of w searches using random enter points. After that the algorithm determines a neighborhood u which contains all neighbors of the each found local minima. The set u is sorted in ascending order by distance from the object new_object to be added. After that new_object is connected with first k nearest elements from the set u.

```
Nearest_Neighbor_Add(object: new_object, integer: k, integer: w)
1   SET[object]: localMins ← Multi_Search (new_object, w);
2   SET[object]: u ←∅ ; //neighborhood;
3   foreach object: local_min ∈ localMins do
4     u ← u ∪ local_min.getFriends();
6   sort the set u so to satisfy the condition σ (u[i],
new_object) < σ (u[i+1], new_object)
7   for (I ← 0; i < k; i++) do
8     u[i].connect(new_object);
9     new_object.connect(u[i]);
```

The choice of the parameter k  is not clear, it depends on the space, but it can be evaluated automatically for an unknown space with a distributed algorithm; we are planning to describe it in our next works. Note that as in 4.2 setting w to  a big number is equivalent to an exhaustive search of the closest elements in the structure. More on the choice of w and k see in the next section.

# 6     Test Results and Discussion

*Test Data*
We have implemented the algorithms presented above in order to validate our assumptions about the scalability of the structure and to evaluate its performance. For a test dataset we have used:

- Uniformly distributed random points with a $L_2$ (Euclidean distance) proximity function (up to $10^6$ elements).
- To test our algorithm in a general metric space we have used a database of chemical compounds [33] with a Tanimoto [34] distance function. We have randomly selected $10^5$ elements from the database to test the algorithm.
- A subset of the TREC-3 documents collection containing 24276 documents[23] for comparison with other works.

*Small World*
To verify the small world properties of the proposed structure we have measured the average path length induced by the greedy search algorithm for the vectors and chemical compounds (see Fig. 3). The plot clearly shows a logarithmic dependence on the dataset size proving it is a navigable small world. Thus the complexity of a single search scales logarithmically. It can be shown that the small world properties retain at any size (we are going to focus on it in one of our next works). Note that for bigger dimensionalities dependence is weaker due to smaller diameter of a set at a fixed number of elements.

*Construction Parameters*
We adjusted the number of search attempts m, so that the probability of finding the true closest element to the query was not less than a fixed value (we took 95% as a reference).

   To test the scaling of the search algorithm with the number of elements $n$ we have plotted (see Fig. 4) the number of multi-searches m required to get the 95% true nearest neighbor rate versus the size of the dataset for $d$=10 and different w parameters of the construction algorithm. For w=20 the dependence is clearly logarithmic up to $10^6$ elements. For low values of w the algorithm complexity dependence deviates from the expected. Arrows denote the point where the dependence deviates from the logarithmic for w=1..4.   One can see that the points are almost equidistant in the logarithmic scale.

   So, if we need to get the logarithmic scaling up to $n$ elements we have to have $w > A \times \log(n)$; (6.1), where $A$ is a constant value. And the overall complexity of both the search and the construction algorithms can be made logarithmic at the same time. Such dependence on the construction parameters can be easily understood. For the low w parameters the probability of finding the true nearest neighbor to a new element is low and the algorithm cannot choose the closest neighbors links correctly. The number of searches required to get $P$ close to unity scales logarithmically with the size of the dataset, leading to equation (6.1). We can set w high enough for any reasonable size of the dataset (like $10^{100}$) while keeping acceptable construction complexity or if the size of the dataset is known (or evaluated dynamically) we can always set the parameters optimal and maintain an overall logarithmic scaling.

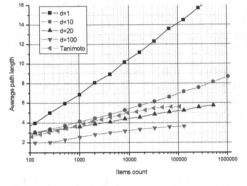 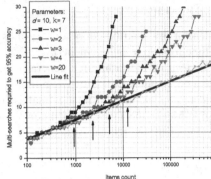

**Fig. 3.** The average hop count induced by a greedy search algorithm for different dimensionality Euclid data and for a chemical compounds dataset (k=10, w=20). The navigable small world properties are evident from the logarithmic scaling.

**Fig. 4.** The number of multi-searches required to get the 95% true nearest neighbor rate versus the size of the dataset for different w parameters of the construction algorithm. Arrows denote points where the dependence deviates from the logarithmic. The points are almost equidistant in the log scale.

Fig. 5 presents the number of multi-searches m for the same parameters as in Fig. 4 setting $w = \lceil A \times \log(n) - c \rceil$ for A=1.5, 2, 2.5, 3. For any value of A the scaling stays logarithmic but at expense of worse complexities for the small values of A. Setting A higher than 2.5 does not affect the complexity of the search.

To define the best choice of the parameter k we have plotted the probability of failing finding the true nearest neighbor versus the fraction of visited elements (metric calculations) for d=10 and different parameters k (see Fig. 6). For k smaller than 2...3·d there is a significant fall of performance, while for bigger values of k there is a very slow decay with the rise of the parameter. For d=2...50 it was verified that the optimal value for k is close to $3 \cdot d$. Also one can see that the probability of a wrong NN result falls exponentially with the fraction of visited elements confirming assumptions from section 4.

The bottom line is that the optimal value for k is $3 \cdot d$; the value of w has to be dynamically changed $A \times \log(n_{current})$ with a constant A or set fixed to $A \times \log(n)$, where n is the maximum database size.

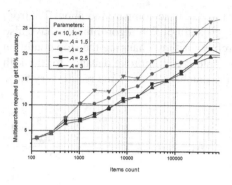
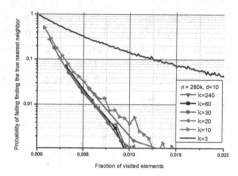

**Fig. 5.** The number of multi-searches required to get the 95% true nearest neighbor rate versus the size of the dataset for a logarithmic scaling of w ($A$=1.5, 2, 2.5, 3)

**Fig. 6.** Probability of failing finding the true nearest neighbor versus fraction of visited elements for $d$=10, $n = 2.6 \cdot 10^5$

### Absolute Speedup and Scaling

The graph (Fig. 7) shows the percent of visited (extracted) elements (vertical) versus the dataset size (horizontal) in a log-log scale for different dimensionalities. k was fixed to $3 \cdot d$ for all trials and w was fixed to a big number. The plot shows that with the increase of the number of elements in the structure, the percentage of visited elements decreases, and the curves become close to straight lines with an angle of 45 degrees (corresponding to the $1/n$ law of decay). This means that the single search complexity does not change significantly with the size of the dataset. From the graphs the overall scaling for complexity of the search can be extracted. It turns out that it scales as $\log^2(n)$, just as it might be expected. One "log" coming from the average path length and the other is from the number of multi-searches.

We have also plotted the average fraction of visited elements for n=$2.6 \cdot 10^5$ in a log-log scale to check the dimensionality dependence (see Fig. 8), it can be approximated by a $d^{1.7}$ power law. Judging on the Fig. 7 it seems that for low $d$ with rise of $n$ at some size the difference in performance between the dimensionalities diminishes. It might be suspected that such behavior will be the same for bigger dimensionalities but it requires further study.

Overall, the measured search complexity scaling for $n>10^5$ and $d = 5..100$ is not worse than $d^{1.7} \times \ln^2(n) \times \ln(1/P_{fail})$ and the construction complexity (deduced from the search complexity) is $d^{1.7} \times n \ln^2(n)$, where $P_{fail}$ is an acceptable probability of failing finding the true nearest neighbor.

To get an idea about how the algorithm performs compared to the other k-NN algorithms we have run a test from [23], a subset of collection TREC-3 documents containing 24276 documents. For a k-NN algorithm we have used a part of the construction algorithm from section 5. To get the averaged 90% recall of 9 documents from the database it required visiting 5% of the database compared to about 2% from the [23]. We believe it is a good result for such a simplified algorithm. We have also made a slight modification of the k-NN search algorithm, changing the stop condition (the algorithm continues to travel the graph while it can improve distance for the k-th element) yielding about 2.5% extraction of the database at the same recall, very close to the state of art.

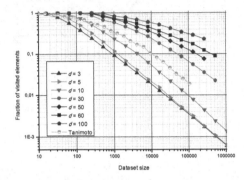
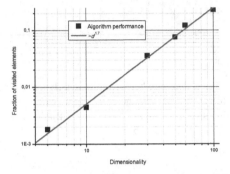

**Fig. 7.** Average fraction of visited elements within a single NN-search versus the size of the dataset for different dimensionality and data types

**Fig. 8.** Average fraction of visited elements within a single Nearest Neighbor search versus the dimensionality of the dataset for $n$ = 262k with a power-law fit

## 7      Conclusions and Future Work

We have proposed a method of organizing data into a distributed small world graph structure suited for the distributed approximate nearest neighbor search in a metric space. The algorithm uses no information about inner topology of the data and space, thus it is applicable to arbitrary metric data. The algorithm is very simple and easy to understand. All elements in the structure are of the same type, there is no central or root element. There is no dedicated algorithm for managing the small world properties, they arise automatically. The algorithm uses only local information on each step and can be initiated from any vertex. The search is approximate from the topological point of view. An unsuccessful Nearest Neighbor query typically results in the second nearest element.

Accuracy of the approximate search can be tuned by using multiple searches with a random initial vertex and the probability of finding a false nearest neighbor decreases exponentially with the number of multi-searches.

The performed simulation for data in the Euclidian space shows that the structure built using the proposed algorithm has the navigable small world property. Both logarithmic search and construction complexity at fixed accuracy can be achieved with appropriate algorithm parameters. There are reasons to believe such behavior will be retained for any dataset size. The algorithm also shows a power law scalability of metric calculation count with dimensionality of the stored data. Simulations for chemical compounds and documents have shown the effectiveness of the approach for non-Euclidian spaces comparable to best algorithms.

The proposed structure was intentionally slimmed-down to demonstrate its scalability over the dataset size and dimensionality. There are several ways to optimize the structure in order to get lower complexity or/and better accuracy constants, such as:

- More complicated algorithms for node friends selection (see sec. 5). It is obvious that selecting nearest neighbors as friends is not the best way to approximate Delaunay graph since it takes into account only distances between the new element and candidates and neglects distances between the candidates. Knowledge of internal structure of the metric space can boost up search performance. In [31] is was shown that for Euclidean space the accuracy of a single search can be significantly increase while keeping the number of friends per node fixed.
- More complicated search algorithms can be used. Excluding already visited elements in consequent searches or/and changing the stop parameters in search algorithm can potentially reduce the number of metric computations several times at the same accuracy.
- More complicated algorithms for navigable small world creation suitable for correlated (non-random) data.

As a future work, we are going to enhance the performance of the structure while keeping good scalability and distributed nature and to make a detail comparison with the state of art algorithms from the area.

Summing up, simplicity, high scalability both with size and data dimensionality and the distributed nature of the algorithm are a good base for building many real-world extreme dataset size high dimensionality similarity search applications.

# References

1. Cover, T.M., Hart, P.E.: Nearest neighbor pattern classification. IEEE Transactions on Information Theory 13(1), 21–27 (1967)
2. Flickner, M., et al.: Query by image and video content: the QBIC system. Computer 28(9), 23–32 (1995)
3. Cost, S., Salzberg, S.: A Weighted Nearest Neighbor Algorithm for Learning with Symbolic Features. Machine Learning 10(1), 57–78 (1993)
4. Sarwar, B., Karypis, G., Konstan, J., Riedl, J.: Item-based collaborative filtering recommendation algorithms. In: Proceedings of the 10th International Conference on World Wide Web, New York, USA, pp. 285–295 (2001)
5. Rhoads, R., Rychlik, W.: A computer program for choosing optimal oligonudeotides for filter hybridization, sequencing and in vitro amplification of DNA. Nucletic Acids Research 17(21), 8543–8551 (1989)
6. Deerwester, S., et al.: Indexing by Latent Semantic Analysis. J. Amer. Soc. Inform. Sci. 41, 391–407 (1990)
7. Kleinberg, J.: The Small-World Phenomenon: An Algorithmic Perspective. In: Annual ACM Symposium on Theory of Computing, vol. 32, pp. 163–170 (2000)

8. Aurenhammer, F.: Voronoi diagrams — a survey of a fundamental geometric data structure. ACM Computing Surveys (CSUR) 23(3), 345–405 (1991)
9. Navarro, G.: Searching in metric spaces by spatial approximation. Paper Presented at the String Processing and Information Retrieval Symposium, Cancun, Mexico
10. Bentley, J.L.: Multidimensional binary search trees used for associative searching. Communications of the ACM 18(9), 509–517 (1975)
11. Finkel, R.A., Bentley, J.L.: Quad Trees: A Data Structure for Retrieval on Composite Keys. Acta Informatica 4(1), 1–9 (1974)
12. Lee, D.T., Wong, C.K.: Worst-case analysis for region and partial region searches in multidimensional binary search trees and balanced quad trees. Acta Informatica 9(1), 23–29 (1977)
13. Samet, H.: The design and analysis of spatial data structures. Addison-Wesley Pub. (1989)
14. Arya, S.: Accounting for boundary effects in nearest-neighbor searching. Discrete & Computational Geometry 16(2), 155–176 (1996)
15. Chávez, E., et al.: Searching in metric space. Journal ACM Computing Surveys (CSUR) 33(3), 273–321 (2001)
16. Arya, S., Mount, D.: Approximate nearest neighbor queries in fixed dimensions. In: SODA 1993 Proceedings of the Fourth Annual ACM-SIAM Symposium on Discrete Algorithms, Philadelphia, PA, USA, pp. 271–280 (1993)
17. Kleinberg, J.: Two algorithms for nearest-neighbor search in high dimensions. In: Proceedings of the Twenty-ninth Annual ACM Symposium on Theory of Computing, STOC 1997, New York, USA, pp. 599–608 (1997)
18. Indyk, P., Motwani, R.: Approximate Nearest Neighbors: Towards Removing the Curse of Dimensionality. In: Proceedings of the Thirtieth Annual ACM Symposium on Theory of Computing, STOC 1998, New York, USA, pp. 604–613 (1998)
19. Kushilevitz, E., Ostrovsky, R., Rabani, Y.: Efficient search for approximate nearest neighbor in high dimensional spaces. In: Proceedings of the Thirtieth Annual ACM Symposium on Theory of Computing, STOC 1998, New York, USA, pp. 614–623 (1998)
20. Gionis, A., Indyk, P., Motwani, R.: Similarity Search in High Dimensions via Hashing. In: Proceedings of the 25th International Conference on Very Large Data Bases, VLDB 1999, San Francisco, USA, pp. 518–529 (1999)
21. Andoni, A., Indyk, P.: Near-Optimal Hashing Algorithms for Approximate Nearest Neighbor in High Dimensions. In: Proceedings of 47th Annual IEEE Symposium on Foundations of Computer Science (FOCS 2006), Berkeley, USA, pp. 459–468 (2006)
22. Houle, M.E., Sakuma, J.: Fast Approximate Similarity Search in Extremely High-Dimensional Data Sets. In: ICDE 2005 (2005)
23. Chávez, E., Figueroa, K., Navarro, G.: Effective Proximity Retrieval by Ordering Permutations. IEEE Transactions on Pattern Analysis and Machine Intelligence 30(9), 1647–1658 (2008)
24. Cai, M., Frank, M., Chen, J., Szekely, P.: MAAN: A Multi-Attribute Addressable Network for Grid Information Services. Journal of Grid Computing 2(1), 3–14 (2004)
25. Ganesan, P., Yang, B., Garcia-Molina, H.: One torus to rule them all: multi-dimensional queries in P2P systems. In: Proceedings of the 7th International Workshop on the Web and Databases, New York, USA, pp. 19–24 (2004)
26. Bharambe, A.R., Agrawal, M., Seshan, S.: Mercury: supporting scalable multi-attribute range queries. In: Proceedings of Applications, Technologies, Architectures, and Protocols for Computer Communication, New York, USA, pp. 353–366 (2004)

27. Beaumont, O., Kermarrec, A.-M., Marchal, L., Riviere, E.: VoroNet: A scalable object network based on Voronoi tessellations. In: Proceedings of International Parallel and Distributed Processing Symposium, Long Beach, US, p. 20 (2007)
28. Novak, D., Zezula, P.: M-Chord: A Scalable Distributed Similarity Search Structure. In: Proceedings of the 2001 Conference on Applications, Technologies, Architectures, and Protocols for Computer Communications, San Diego, pp. 149–160 (2001)
29. Batko, M., Gennaro, C., Zezula, P.: Similarity Grid for Searching in Metric Spaces. In: Türker, C., Agosti, M., Schek, H.-J. (eds.) Peer-to-Peer, Grid, and Service-Orientation in Digital Library Architectures. LNCS, vol. 3664, pp. 25–44. Springer, Heidelberg (2005)
30. Haghani, P., Michel, S., Aberer, K.: Distributed similarity search in high dimensions using locality sensitive hashing. Paper presented at the 12th International Conference on Extending Database Technology: Advances in Database Technology, New York, USA
31. Beaumont, O., Kermarrec, A.-M., Rivière, É.: Peer to peer multidimensional overlays: approximating complex structures. In: Proceedings of the 11th International Conference on Principles of Distributed Systems, Berlin, Heidelberg (2007)
32. Krylov, V., Ponomarenko, A., Logvinov, A., Ponomarev, D.: Single-attribute Distributed Metrized Small World Data Structure. Paper Presented at the IEEE International Conference on Intelligent Computing and Intelligent Systems (CAS)
33. Wang, Y., Xiao, J., Suzek, T.O., Zhang, J., Wang, J., Bryant, S.H.: PubChem: a public information system for analyzing bioactivities of small molecules. Nucl. Acids Res. 37, W623–W633 (2009)
34. James, C.A., Weininger, D., Delaney, J.: Fingerprints-Screening and Similarity (1997), http://www.daylight.com/dayhtml/doc/theory/theory.toc.html

# Parallel Approaches to Permutation-Based Indexing Using Inverted Files

Hisham Mohamed and Stéphane Marchand-Maillet

Université de Genève, Geneva, Switzerland
{hisham.mohamed,stephane.marchand-maillet}@unige.ch

**Abstract.** We present parallel strategies for indexing and searching permutation-based indexes for high dimensional data using inverted files. In this paper, three strategies for parallelization are discussed; posting lists decomposition, reference points decomposition, and multiple independent inverted files. We study performance, efficiency, and effectiveness of our strategies on high dimensional datasets of millions of images. Experimental results show a good performance compared to the sequential version with the same efficiency and effectiveness.

**Keywords:** Approximate similarity search, Large scale distributed indexing, Permutation based indexes, Parallel metric inverted files.

## 1 Introduction

The way of answering users' queries depends on the search scenario. The "exact match" scenario is commonly used, where the system retrieves all matches to a given query from the database. Nowadays, this way of answering the query is not the most useful for some applications such as text plagiarism to track the similarity between an article against a database of texts, multiple genome comparison to find all the similarities between one or more genes, and multimedia retrieval to find the most similar picture or video to a given example. The similarity search paradigm [1] is more applicable on these models.

For a query $q$ and a data collection $D$, similarity search sorts all the data items by similarity to the given query according to a given distance function $d : D \times D \longrightarrow \Re$. The most relevant objects to the query are the k-top ranked objects (k-NN query) or the objects located within a distance range $\rho$ from the query (range query). Several techniques have been developed for improving the performance of the similarity search problem [2]. One of the research topics still attracting interest is the scalability of similarity search for high-dimensional data. Different approaches have been proposed to attack the curse of dimensionality [3]. One of the most promising routes is the approximate similarity search [4,5]. It proposes solutions to improve the performance when handling high dimensional data at the price of effectiveness.

*Permutation-based indexes* are the most recent technique for approximate similarity search [6,7]. In this work, we propose several distributed algorithms for handling *permutation-based indexes* using inverted files. We describe our ideas through three

G. Navarro and V. Pestov (Eds.): SISAP 2012, LNCS 7404, pp. 148–161, 2012.

levels of parallelization and have tested them on high dimensional datasets, which consist of 4,594,734 million objects. The rest of the paper is organized as follows. In the next section, a brief background about *permutation-based indexes* is given. Section 3 gives a review of the related work. In section 4, we detail the main ideas about our parallel strategies. Then, we present our results in section 5 and conclude in section 6.

## 2    Basic Notations and Definitions

The intuition behind the *permutation-based indexes* is based on "predicting closeness between elements according to how they order their distances towards a distinguished set of anchor objects" [6,7]. Given a collection of $N$ objects $o_i$ in a domain $D = \{o_1 \ldots o_N\}$, and a distance function $d : D \times D \to \Re$ between the objects. We assume that the distance function $d(.,.)$ follows the metric space postulates [2] $\forall o_i, o_j, o_k \in D$:

- $o_i = o_j \iff d(o_i, o_j) = 0$ *identity*,
- $d(o_i, o_j) \geq 0$ *non-negativity*,
- $d(o_i, o_j) = d(o_j, o_i)$ *symmetry* and
- $d(o_i, o_k) \leq d(o_i, o_j) + d(o_j, o_k)$ *triangle inequality*.

A set of $n$ reference objects $R = \{r_0, r_1, \ldots r_{n-1}\} \subset D$ is randomly selected from $D$. Each object $o_i \in D$ is represented by an ordered list $L_{o_i}$. The ordered list for each object contains the reference points set sorted by their distance $d$ to the object $o_i$. More formally, $L_{o_i}$ is the permutation of $(0, \ldots, j, \ldots, n-1)$ according to the distance function $d$. $P(L_{o_i}, r_j)$ returns the position of the reference object $r_j$ within the ordered list $L_{o_i}$ of object $o_i$. For example, $P(L_{o_i}, r_j) = 5$ means that $r_j$ is the 5th nearest reference point to the object $o_i$. Figures 1a and 1b show a group of objects and their ordered list respectively.

The permutation lists for all object are saved in the main memory. For a given query $q$, an ordered list $L_q$ is computed as for the database objects with respect to the same reference points. The similarity between the query and database objects is measured by comparing the permutation lists using *Spearman Footrule Distance(SFD)*[2].

$$SFD(o_i, q) = \sum_{r \in R} |P(L_{o_i}, r) - P(L_q, r)| \tag{1}$$

## 3    Related Work

Several works were proposed to speed up *permutation-based indexes* using various techniques. Amato and Savino [7] proposed an algorithm to store the permutations using inverted files. Figuerroa et al. [8] speeded up the distance calculation between the objects by indexing the permutation relative to their distance from the reference points. In [9], authors proposed the *brief permutation index* which is a technique to reduce memory usage and speed up the distance calculation. The main idea is to encode the

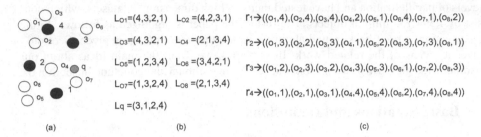

$L_{o1}=(4,3,2,1)$  $L_{o2}=(4,2,3,1)$  $r_1 \rightarrow ((o_1,4),(o_2,4),(o_3,4),(o_4,2),(o_5,1),(o_6,4),(o_7,1),(o_8,2))$

$L_{o3}=(4,3,2,1)$  $L_{o4}=(2,1,3,4)$  $r_2 \rightarrow ((o_1,3),(o_2,2),(o_3,3),(o_4,1),(o_5,2),(o_6,3),(o_7,3),(o_8,1))$

$L_{o5}=(1,2,3,4)$  $L_{o6}=(3,4,2,1)$  $r_3 \rightarrow ((o_1,2),(o_2,3),(o_3,2),(o_4,3),(o_5,3),(o_6,1),(o_7,2),(o_8,3))$

$L_{o7}=(1,3,2,4)$  $L_{o8}=(2,1,3,4)$

$L_q=(3,1,2,4)$     $r_4 \rightarrow ((o_1,1),(o_2,1),(o_3,1),(o_4,4),(o_5,4),(o_6,2),(o_7,4),(o_8,4))$

(a)                        (b)                                      (c)

**Fig. 1.** Example of Metric inverted files a) Black dots are reference objects; white dots are data objects; the grey dots is query object b) ordered lists for all data objects $o_i$. c)Inverted index; the vocabulary are the reference points and the posting lists are pairs of data objects and their positions.

permutation as a binary vector and to compare these vectors using Hamming distance. The most recent work which is based on *permutation-based indexes* is the *prefix permutation index* (PP-Index) [10,11]. The PP-Index stores the prefix of the permutations only and measure the similarity between objects based on the length of its shared prefix. Some techniques for parallelization are also proposed for speeding up the PP-Index.

Other work which is based on measuring similarity based on number reference points is the work done in [12]. Novak and Batko [12] proposed the M-Index algorithm. M-Index maps the objects to a numeric domain. This is done by selecting number of pivots; reference points, which represent the object then the distance functions $d(.,.)$ between the objects and the reference points is normalized by a constant value which is greater than the maximum distance $d$ between any two objects in the data domain. The M-Index data structure provides exact and approximate similarity search.

### 3.1 Metric Inverted Files

Our work is based on the metric inverted files (MIF) which were proposed by Amato and Savino [7]. Inverted files[13,14] are mainly used for text indexing. An inverted file consists of two parts, the *vocabulary* and the *posting list*. The vocabulary is the list of all unique terms and the posting list is associated to every vocabulary words and stores all the locations of the vocabulary word in the corpus. In MIF, the vocabulary is the set of reference points and the posting list for a reference point $r_j$ contains a list of pairs $(o_i, P(L_{o_i}, r_j)) \forall o_i \in D$. Figure 1c shows an example of MIF. Algorithm 1 [7] explains the main idea for searching for a given query $q$. An accumulator is assigned to each object $o_i \in D$ and initialized to zero. The posting list for each reference point is accessed and the accumulator is updated by adding the difference between the position of the current reference object $r_j$ in the ordered list of the query and the $P(L_{o_i}, r_j)$ of the objects in the posting list, using equation (1). After checking the posting lists of all the reference points, the objects are sorted based on their accumulator value. Objects with small accumulator value are more similar to the query object.

**Algorithm 1**

*IN: Query: q,*
  *Reference Object list of n elements: R,*
  *Posting lists assigned to each reference object for N objects;*
*OUT: Sorted Objects list: out*
1.   *Create a list of accumulators $A[1 \ldots N]$*
2.   *Set accumulators values to 0*
3.   *For each $r \in R$*
4.     *Let $\Delta$ be the posting list for the reference object $r$*
5.     *Set $i \longleftarrow 0$*
6.     *For each pair $(o, P(L_o, r)) \in \Delta$*
7.       *Set $A[i] = A[i] + |P(L_o, r) - P(L_q, r)|$*
8.         *$i \longleftarrow i + 1$*
9.   *Sort(A)*
10.  *out $\leftarrow$ A*

In [7], authors have improved the performance of the algorithm by indexing the objects with respect to some nearest reference objects only and perform the search using these nearest reference objects. They experimentally proved that the nearest reference objects are the most relevant ones. The complexity of this basic algorithm is $O(nN)$, where $n$ is the number of reference objects and $N$ is the number of objects. Our parallel algorithm is based on this basic algorithm and the modification which was proposed in [7] can be applied easily to our parallel models. We present our parallel strategy through three levels of parallelization as shown in the next section. Our first and second parallel strategies follow the parallel inverted files for text indexing in [15], but we apply it for approximate similarity metric searching.

## 4  Parallel Metric Inverted Files

Our parallelization strategy works on different levels of decreasing complexity and increasing throughput: 1) *posting lists decomposition*, 2) *reference points decomposition*, and 3) *multiple independent inverted files*.

### 4.1  Posting Lists Decomposition (PLD)

We first present the basic way of parallelization, which is based on posting lists decomposition. We partition the data equally and build the inverted files with global reference points.

**Indexing.** The data domain $D$ of $N$ objects is randomly divided into sub-domains of equal sizes $D_0 \ldots D_p$, where $p$ is the number of parallel processes. Hence, every process $i$ nominates $n_i$ reference points from its partial data. Since the data is divided, the nominated reference points for each process are unique and not known by the other

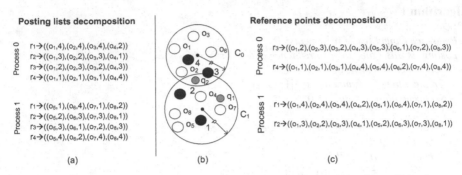

**Fig. 2.** a) Posting lists decomposition algorithm. b) Data clustering: The data within the same circle are more related to each other. Each cluster has center, radius $\rho$ and two reference points which are the black dots. c) Reference points decomposition algorithm.

processes. Then, all to all communication is done between the processes to share the reference points information. Each process then starts to build its own inverted file data structure based on the global reference points and the partial data it has access to. Accordingly, each process is responsible for all the references with partial posting list. Figure 2a shows posting list decomposition. For example, the posting list for reference point 1 in figure 1c is divided into two partial lists. Each partial list is processed by a different process.

**Searching.** The inverted file is partitioned. Therefore, to answer a query $q$, all partial inverted files need to be scanned. A broker process accepts query requests. These queries are then broadcasted to all other processes. After receiving, each process starts to index the query and apply the search on its local inverted file. Once done, every process sends its local accumulators to the broker process. The broker concatenates the accumulators and sorts the objects based on their accumulator values. More formally, in Algorithm 1, the **for** loop in lines 6-8 can run in parallel over the different partial domains $D_0, D_1, \ldots D_p$. Thus, theoretically the memory usage is reduced from $O(nN)$ to $O(n\frac{N}{p})$ and the complexity leads to $O(n\frac{N}{p}) + t_s$, where $n$ is the number of the reference points and $t_s$ is the time needed to receive the accumulators of the partial ranked objects. In this algorithm, all the reference points need to be checked and all the processes have to participate to answer a query. To improve this, we introduce two additional strategies for parallelization.

### 4.2 Reference Points Decomposition (RPD)

Amato et al. in [7] experimentally proved that a high number of reference points does not really improve the accuracy of the method and have suggested to apply search using the nearest objects only. We use this evidence in order to improve our parallel strategy. In this strategy, the inverted index is divided based on the number of reference points.

**Indexing.** Similarly to section 4.1, the data domain is divided into partial domains, but not randomly. We cluster that data into $p$ clusters, based on the number of processes. For a cluster $i$ a center $c_i$ and a radius $\rho_i$ are to be determined. The center $c_i$ is an average object from other objects in the cluster and its radius $\rho_i$ is the distance from the center to the farthest object in the cluster. Then, each process $i$ nominates a fixed number $n_p$ of reference points from its cluster and share it with all other processes. Each process indexes its data based on these reference points. Hence, we obtain the same inverted file distribution as in section 4.1. Afterwards, each process asks all other processes about its partial posting list which is related to its local reference points. The processes combine the partial received posting lists. When exchanging is done, each process deletes the other reference points with their partial posting lists as there is no more need for them. Now, every process $i$ is responsible for $n_p$ reference points with their posting lists, which is related to the whole dataset. Only the broker process needs to know the total reference lists without the posting lists for searching as explained next. Figures 2b and 2c show examples of clustered data and the reference points decomposition structure.

**Searching.** Before accepting queries, each process $i$ sends the center point $c_i$ and the radius $\rho_i$ of its cluster to the broker process. Once a query $q$ is sent to the broker process, it computes the distance between the query and the centers of the clusters. If the query is located within the region of a certain cluster, then this cluster is considered as a searching region. If not, the cluster is neglected. This can be calculated using this equation:

$$d(q, c_i) < \rho_i \qquad (2)$$

The broker generates the ordered list of the query object $L_q$ with respect to the nearest reference points defined by the cluster within which the query is located. For example, in figure 2b, query point $q_1$ is within cluster 1, so the ordered list is created with respect to the reference points of cluster 1 and the ordered list is $(1, 2)$. For query point $q_2$, it is located between the two clusters, so the ordered list is $(3, 2, 4, 1)$.

Once the ordered list $L_q$ is created, the broker sends it to the responsible processes $p$. Each process ranks the local objects exactly like algorithm 1, but with respect to their local reference points. The broker then gathers all the accumulator lists from the responsible processes and sum the accumulators for the same objects. Finally it sorts the accumulator list and sends the ordered object to the user. Algorithm 2 shows the searching process for reference points decomposition algorithm. Lines 2-4 are executed by the broker process to define the active clusters and to generate the query ordered list. Lines 6-12 are similar to algorithm 1, but each active process checks its local reference points only instead of checking all the reference points. Lines 14-21 are executed by the broker process to gather, sum and sort the accumulators. Theoretically the memory usage becomes $O(n_p N)$, where $N$ is the number of objects. The complexity becomes $O(n_p N) + t_r$, where $t_r$ is the time required to receive $pN$ lists. At the same time, we have increased the throughput of the system since more than one query can be answered simultaneously if they are located in different clusters. The main drawback of this algorithm is the increase of the complexity in the construction of the inverted files as the data needs to be exchanged between all the processes.

**Algorithm 2**

*IN:    Query: q,*
*Global reference objects list of n objects: globalR,*
*Local reference objects list on $n_p$ elements: localR,*
*Posting lists assigned to each local reference object for N objects,*
*Cluster list: CL;*

*OUT: Sorted Objects list: out*

1.    *if(broker)*
2.        *Search_for_responsible_clusters(q,CL)*
3.        *Create empty ordered list for q: $L_q \leftarrow \phi$*
4.        *Generate_query_ordered_list(CL,globalR,$L_q$)*
5.    *if(active)*
6.        *Create l accumulators lists based on the number of active processes: $a_l \leftarrow 0$*
7.        *For each $r \in localR$*
8.            *Let $\Delta$ be the posting list for the reference object r*
9.            *Set $i \longleftarrow 0$*
10.           *For each pair $(o, P(L_o, r)) \in \Delta$*
11.               *Set $a_l[i] = a_l[i] + |P(L_o,r) - P(L_q,r)|$*
12.               *$i \longleftarrow i+1$*
13.   *if(broker)*
14.       *Create accumulator list of N: A*
15.       *For each partial $a_l$ in CL*
16.           *Set $i \longleftarrow 0$*
17.           *For each object o in $a_l$*
18.               *A[i]=A[i]+$a_l[i]$*
19.               *$i \longleftarrow i+1$*
20.   *Sort(A)*
21.   *out $\leftarrow$ A*

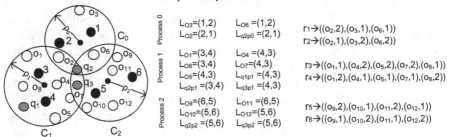

**Multiple independent inverted files**

**Fig. 3.** Multiple inverted files: $p$ independent inverted files are created. Each inverted file represents a cluster and is assigned to a certain process.

### 4.3   Multiple Independent Inverted Files (MIIF)

In [7], authors used $k_i$ nearest reference points to index the data. We map this technique onto our parallel algorithm. Thus, we reduce the range of the search using partial data with partial reference points.

**Indexing.** Similar to section 4.2, the data domain is clustered based on the number of processes. Every cluster nominates a number of reference points near to its center, but in this algorithm, clusters do not share the reference points with other clusters. Each process builds its own inverted file, hence there are $p + 1$ inverted files $\Pi_0, \ldots \Pi_p$. Each inverted file represents different data with respect to different reference points. Processes only share the information about their centers and their radiuses. Figure 3 shows example of multiple inverted files.

**Searching.** By sharing the centers and radiuses information between all the processes, the query can be submitted to any process (any process becomes a broker). Once a query is submitted to a process $i$, the process recognizes the responsible clusters similar to 4.2 using equation 2 and sends the query point information to the corresponding responsible processes. Each process indexes the data based on its local reference points and ranks the objects similar to algorithm 1. Hence, we have $l$ output lists for different objects sorted based on different reference points. To combine and rank them, each process computes the distance between top ranked *2K-points* and the query. Then, these distances are sent to the process which received the query. Finally, the process ranks these objects based on their distances and sends the results back to the user. For example, for *NNquery(q,30)* the algorithm returns the most similar $K = 30$ objects to the query. Each process searches within its local inverted index data structure and sends to the master process the distance between the query and the first $2K = 60$ objects. So, if the query is located within 2 clusters, the master node receives $2 \times 2K = 120$ distances. These 120 distances are sorted and the top 30 objects are sent back to the user. We empirically choose the first *2K* in order to improve the results of the system. Algorithm 3 shows the search process for MIIF algorithm. Lines 2-3 are executed by the process which receives the query (master). Lines 5-12 compute the accumulators with respect to the local data and calculate the distance between the query and the top $2K$ objects. Lines 14-16 are used to combine the result lists. This theoretically leads to $O(n_p N_p) + t_m$ query time and the memory usage becomes $O(n_p N_p)$, where $N_p$ is the maximum number of objects located within a cluster with respect to all other clusters, $n_p$ is the number of reference objects within a cluster and $t_m$ is time needed to receive $2K$ calculated distances from $p$ processes where $p$ is the number of clusters intersecting with the query(number of active processes). In summary, table 1 shows the complexity and the memory occupation of the three algorithms.

**Table 1.** Complexity and searching time for the sequential and parallel algorithms. $n$: Number of reference points. $n_p$: Number of reference points per process $p$. $N$: Number of objects. $N_p$: Maximum number of objects within a cluster with respect to other clusters. $t_s$: Communication time to receive $N$ objects. $t_r$: Communication time to receive $pN$ objects. $t_m$: Communication time to receive $2Kp$ distances.

| Algorithm | Memory occupation | Complexity |
|---|---|---|
| Sequential MIF | $O(nN)$ | $O(nN)$ |
| PLD | $O(n\frac{N}{p})$ | $O(n\frac{N}{p}) + t_s$ |
| RPD | $O(n_p N)$ | $O(n_p N) + t_r$ |
| MIIF | $O(n_p N_p)$ | $O(n_p N_p) + t_m$ |

**Algorithm 3**

*IN:    Query: q,*
   *Number of top ranked queries $K$,*
   *Local reference objects list on $n$ elements: $localR$,*
   *Posting lists assigned to each local reference object for $N_p$ objects,*
   *Cluster list: $CL$;*
*OUT: Sorted Objects list: out*

1. *if(master)*
2.  *Search_for_responsible_clusters(q,CL)*
3.  *Create $l$ accumulators lists based on the number of active processes: $a_l \leftarrow 0$*
4. *if(active)*
5.  *For each $r \in localR$*
6.   *Let $\Delta$ be the posting list for the reference object $r$*
7.   *Set $i \leftarrow 0$*
8.   *For each pair $(o, P(L_o, r)) \in \Delta$*
9.    *Set $a_l[i] = a_l[i] + |P(L_o, r) - P(L_q, r)|$*
10.    *$i \leftarrow i + 1$*
11.  *Create $l$ distance lists: $d_l \leftarrow 0$*
12.  *Calculate_distance(q, 2k, A_l, d_l)*
13. *if(master)*
14.  *Combine the partial distance lists into $D$*
15.  *Sort(D)*
16.  *out $\leftarrow D$*

## 5  Experimental Results

We have conducted large-scale experiments to test the validity of our models. We have implemented the sequential and the three parallel algorithms using C++ and MPICH2. For data clustering, we used the *k-means* algorithm presented in [16]. Then, we used our implementation to index 4,594,734 (84-dimensional) color features related to 4,594,734 images from the 12-million ImageNet corpus [17]. MPICH2 is installed on a Linux cluster of 20 DualCore computers (40 cores in total) holding each 8Gb of memory and 512Gb of local disk storage, led by a master 8-core computer holding 32Gb of memory and a TeraByte storage capacity.

We measured the running time and the effect of increasing the number of cores. Also, we measure both the recall and the position error [2] for each algorithm. Given a query $q$ the recall is defined as:

$$Recall = \frac{|S \cap S_A|}{|S|} \tag{3}$$

and the position error is defined as:

$$Position\ Error = \frac{\sum_{o \in S_A} |P(X, o) - P(S_A, o)|}{|S_A| \cdot |D|} \tag{4}$$

where $S$ and $S_A$ are the ordering of $K$ top ranked objects to $q$ for exact similarity search and approximate similarity search respectively. $X$ is the ordering of dataset $D$ with respect to their distance from $q$ and $P$ is defined in section 2. In all of our experiments, we measure the average value based on 10 different queries from the datasets.

## 5.1   Indexing

Due to memory limitations, the sequential algorithm can not handle this data. We have 4,594,734 objects and each pair in the posting list needs about 8 bytes. For 1,000 reference objects we need about 34GB of memory, which cannot be supported individually by any of our machines.

Figures 4a, 4b and 5 show the indexing time for algorithms presented in sections 4.1, 4.2 and 4.3 respectively. The x-axis shows the number of cores used for indexing and the y-axis shows the running time in seconds. The running time for the clustering of the RPD and the MIIF algorithms are not included in the figures. The clustering time varies from 5 to 7 hours according to the number of clusters. For the three algorithms, when the number of cores increases, the indexing time decreases. As we can see, RPD takes more time than PLD, due to the communication required between the nodes in order to build the inverted files. For both strategies, when the number of reference objects increases the running time increases. Memory overload can happen and stop the indexing process. In our computer cluster, every 2 cores share 8GB of memory. For *PLD* the data is equally divided, so if overload happens at one process, it will happen for all other processes. For *RPD*, this can be done by any process irrespective of the number of reference points as the data are clustered. So, some clusters may have more data than others. The organization of the clusters should be done, so as to avoid large clusters to be located within the same node. In our experiments, this happened with 4,000 reference points as one of the processes had a lot of data in its cluster, and started to write in the swap and affected the indexing time.

For *MIIF*, the data is totally independent, there is no need for exchanging data and the number of reference points does not have a significant effect on the memory usage. Even if the data and the reference points are independent, the memory overload effect can still happen so that careful data organization is required in order not to allocate large clusters together in the same machine.

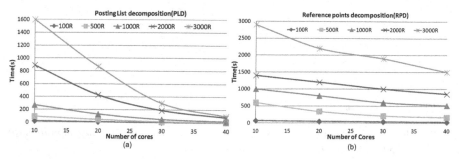

**Fig. 4.** Indexing times for *posting list decomposition* (a) and *reference points decomposition* (b) relative to 100, 500, 1000, 2000 and 3000 reference points (R)

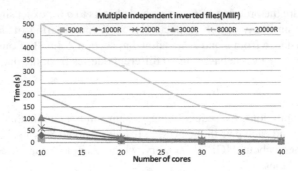

**Fig. 5.** Indexing times for *multiple independent inverted files* relative to 500, 2000, 3000, 8000 and 20000 reference points (R)

## 5.2 Searching

Figures 6a, 6b and 7 show the searching time for algorithms *PLD*, *RPD* and *MIIF* respectively. The x-axis shows the number of cores and the y-axis shows the running time in seconds. Similar to indexing when the number of cores increases the average response time decreases. We found that the average searching time using *PLD* is much faster than *RPD*. The main reason is the communication time needed to gather the results. If the query is located in the range of more than one cluster, $pN$ pairs need to be sent back to the broker process, where $p$ is the number of assigned clusters. For example, if we have 40 clusters and the query is located with in these clusters. Each process sends $N$ pairs to the broker. So, the broker receives $40N$ pairs, which affects the running time, but for *PLD* the broker receives only $N$ pairs. Then, it is normal that the average time of searching using posting lists decompositions is much faster than the reference points decompositions, due to the communication overhead. In *MIIF*, the data received by the broker is $2Kp$, but it is not faster than *RPD* due to the distance calculation between the top $2K$ retrieved points and the query.

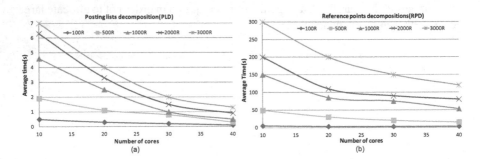

**Fig. 6.** Searching times for *posting list decomposition* (a) and *reference points decomposition* (b) relative to 100, 500, 1000, 2000 and 3000 reference points (R)

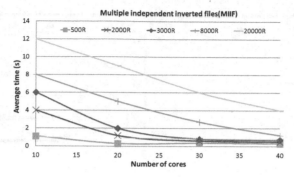

**Fig. 7.** Searching times for *multiple independent inverted files* relative to 500, 2000, 3000, 8000 and 20000 reference points (R)

## 5.3 Recall and Position Error

Figures 8a and 8b show the average recall and the average position error for algorithms *PLD* and *RPD* relative to 1, 10 , 30, 50 $K$ points. The average recall and the average position error for both of our algorithms is similar to those obtained using the sequential implementation [7] with better computing performance, as we have the same inverted files but distributed on multiple nodes with different techniques. Figures 9a and 9b show the average recall and the average position error for algorithm *MIIF* relative to 1, 10 , 30, 50 $K$ points. For high number of reference points, the recall and position error are improved due to the direct distance calculation between the top $2pK$ objects and the query object. For low number of reference points, the effectiveness is degraded because these points are distributed on the clusters and they are independent. For example, for 100 reference points on 40 cores, every process generates 2 reference points related to its data. These numbers of reference points are not enough to distribute the data. The recall is then 0 and position error is 1.8. For 20,000 reference points, each core has 500 reference points and this improve the recall and the position error.

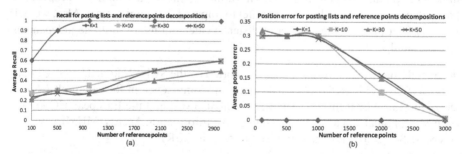

**Fig. 8.** a) Recall for *posting lists decompositions* and *reference points decompositions* algorithms. The x-axis shows the number of reference points and the y-axis shows the average recall b) Position error for the two algorithms. The x-axis shows the number of reference points and the y-axis shows the average position error. Recall and position error measured relative to $K = 1$, 10, 30, 50 points.

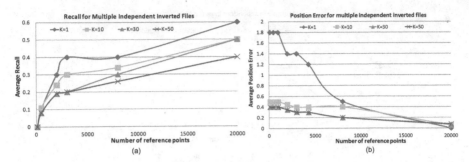

**Fig. 9.** a) Recall for *multiple independent inverted files* algorithm. The x-axis shows the number of reference points and the y-axis shows the average recall. b) Position error for the same algorithm. The x-axis shows the number of reference points and the y-axis shows the average position error. Recall and position error measured relative to $K = 1$, 10, 30, 50 points.

## 6  Conclusion

We have presented three parallel algorithms to index and search permutation-based indexes using inverted files. For the three algorithms, when the number of cores increases, the indexing and searching time decreases. The posting lists decompositions and the reference points decompositions are the same but with different distribution of the data. Both of the algorithms give good recall and position error values with respect to the same number of nodes, but the reference points decomposition algorithm is slower in searching as the broker node needs to gather $pN$ lists in order to rank the objects. The multiple independent inverted files algorithm requires a high number of reference points in order to increase its performance, because the data and the references are independently distributed on the processes. The direct distance calculation is useful to improve the recall and the point error, but it affects the running time.

For future work, we will study how the choice of the references points can affect the results of our parallel models. Also, we will further combine the techniques proposed in [7] to each process independently in parallel to improve the running time. Also, we will study the effect of changing number of distance calculations for multiple independent inverted files algorithm.

**Acknowledgment.** This work is jointly supported by the Swiss National Science Foundation (SNSF) via the Swiss National Center of Competence in Research (NCCR) on Interactive Multimodal Information Management (IM2) and the European COST Action on Multilingual and Multifaceted Interactive Information Access (MUMIA) via the Swiss State Secretariat for Education and Research (SER).

## References

1. Jagadish, H.V., Mendelzon, A.O., Milo, T.: Similarity-based queries. In: Proceedings of the Fourteenth ACM SIGACT-SIGMOD-SIGART Symposium on Principles of Database Systems, PODS 1995, pp. 36–45. ACM, New York (1995)

2. Zezula, P., Amato, G., Dohnal, V., Batko, M.: Similarity Search: The Metric Space Approach. Advances in Database Systems, vol. 32. Springer (2006)

3. Samet, H.: Foundations of multidimensional and metric data structures. The Morgan Kaufmann series in computer graphics and geometric modeling. Elsevier/Morgan Kaufmann (2006)

4. Indyk, P., Motwani, R.: Approximate nearest neighbors: towards removing the curse of dimensionality. In: Proceedings of the Thirtieth Annual ACM Symposium on Theory of Computing, STOC 1998, pp. 604–613. ACM, New York (1998)

5. Patella, M., Ciaccia, P.: Approximate similarity search: A multi-faceted problem. J. of Discrete Algorithms 7(1), 36–48 (2009)

6. Gonzalez, E.C., Figueroa, K., Navarro, G.: Effective proximity retrieval by ordering permutations. IEEE Transactions on Pattern Analysis and Machine Intelligence 30(9), 1647–1658 (2008)

7. Amato, G., Savino, P.: Approximate similarity search in metric spaces using inverted files. In: Proceedings of the 3rd International Conference on Scalable Information Systems, InfoScale 2008, pp. 28:1–28:10. ICST (Institute for Computer Sciences, Social-Informatics and Telecommunications Engineering), ICST (2008)

8. Figueroa, K., Fredriksson, K.: Speeding up permutation based indexing with indexing. In: Proceedings of the 2009 Second International Workshop on Similarity Search and Applications, SISAP 2009, pp. 107–114. IEEE Computer Society, Washington, DC (2009)

9. Téllez, E.S., Chávez, E., Camarena-Ibarrola, A.: A Brief Index for Proximity Searching. In: Bayro-Corrochano, E., Eklundh, J.-O. (eds.) CIARP 2009. LNCS, vol. 5856, pp. 529–536. Springer, Heidelberg (2009)

10. Esuli, A.: Mipai: Using the pp-index to build an efficient and scalable similarity search system. In: Proceedings of the 2009 Second International Workshop on Similarity Search and Applications, SISAP 2009, pp. 146–148. IEEE Computer Society, Washington, DC (2009)

11. Esuli, A.: Pp-index: Using permutation prefixes for efficient and scalable approximate similarity search. In: Proceedings of LSDSIR 2009, vol. i, pp. 1–48 (July 2009)

12. Novak, D., Batko, M., Zezula, P.: Metric index: An efficient and scalable solution for precise and approximate similarity search. Inf. Syst. 36(4), 721–733 (2011)

13. Zobel, J., Moffat, A., Ramamohanarao, K.: Inverted files versus signature files for text indexing. ACM Trans. Database Syst. 23(4), 453–490 (1998)

14. Zobel, J., Moffat, A.: Inverted files for text search engines. ACM Comput. Surv. 38(2) (July 2006)

15. Stanfill, C.: Partitioned posting files: a parallel inverted file structure for information retrieval. In: Proceedings of the 13th Annual International ACM SIGIR Conference on Research and Development in Information Retrieval, SIGIR 1990, pp. 413–428. ACM, New York (1990)

16. Kanungo, T., Mount, D.M., Netanyahu, N.S., Piatko, C.D., Silverman, R., Wu, A.Y.: An efficient k-means clustering algorithm: Analysis and implementation. IEEE Transactions on Pattern Analysis and Machine Intelligence 24, 881–892 (2002)

17. Deng, J., Dong, W., Socher, R., Li, L.J., Li, K., Fei-Fei, L.: ImageNet: A Large-Scale Hierarchical Image Database. In: CVPR 2009 (2009)

# Super-Linear Indices
# for Approximate Dictionary Searching

Leonid Boytsov

11710 Old Georgetown Rd., Unit. 612, North Bethesda, MD, 20852, USA
leo@boytsov.info

**Abstract.** We present experimental analysis of approximate search algorithms that involve indexing of deletion neighborhoods. These methods require huge indices whose sizes grow exponentially with respect to the maximum allowable number of errors $k$. Despite extraordinary space requirements, the super-linear indices are of great interest, because they provide some of the shortest retrieval times.

A straightforward implementation that creates a hash index directly over residual strings (obtained by deletions from dictionary words) is not space efficient. Rather than memorizing complete residual strings, we record only deleted characters and their respective positions. These data are indexed using a perfect hash function computed for a set of residual dictionary strings [2].

We carry out an experimental evaluation of this approach against several well-known benchmarks (including FastSS, which stores residual strings directly [3]). Experiments show that our implementation has a comparable or superior performance to that of the fastest benchmarks. At the same time, our implementation requires 4-8 times less space as compared to FastSS.

**Keywords:** wildcard neighborhood generation, reduced alphabet neighborhood generation, Mor-Fraenkel method, perfect hashing, FastSS.

## 1 Introduction

Approximate string searching is ubiquitous in information retrieval, spellchecking, computational biology, speech recognition, and security software (e.g., for detection of weak passwords). This problem is twofold: finding the locations of a pattern inside a given text, and finding matching strings in a set, i.e., in a dictionary. In both cases, the pattern needs to match data only approximately. A degree of closeness is determined by a distance function. We restrict our attention to the case of lossless methods which guarantee retrieval of all words within the Levenshtein distance $k$ from the search pattern [15]. This distance function is equal to the minimum number of basic edit operations (insertions, deletions, and substitutions) required to convert one string into another. Note that we are primarily interested in practical aspects of this problem and evaluate only those methods that are capable of tolerating more than one error (i.e., support $k > 1$).

G. Navarro and V. Pestov (Eds.): SISAP 2012, LNCS 7404, pp. 162–176, 2012.

We focus on the methods with super-linear indices, which rely on generation of deletion neighborhoods and/or reduced alphabet neighborhoods. In particular, deletion neighborhoods are memorized in the index. The redundancy in storage allows us to achieve very short retrieval times. It is also noteworthy that these methods involve neither direct computation of the Levenshtein distance nor explicit verification if the distance between the pattern and dictionary words is at most $k$.

The rest of the paper is organized as follows. Prior art is described in Subsection 1.1. In Subsection 1.2, we introduce notation and formalize the problem. The implemented methods are described in Section 2, which starts with a discussion on the concepts of full and wildcard neighborhood generation. The experiments are presented in Section 3. Section 4 concludes the paper.

## 1.1  Related Work

Damerau [8] presented misspelling statistics and described one of the first methods of approximate dictionary searching. This method could tolerate only a single error. Levenshtein proposed a string similarity function that is equal to the minimal number of insertions, deletions, and substitutions necessary to make strings equal. A dynamic programming algorithm to efficiently compute Levenshtein distance was independently discovered by several scientists [20]. This classic algorithm has a quadratic complexity and a number of improvements were suggested [18].

To further reduce retrieval time, it is necessary to index the data. There are a lot of indexing techniques for approximate dictionary searching, which rely, among other methods, on generating neighborhoods, indexing of contiguous and gapped string subsequences, as well as on organizing a dictionary in the form of a trie (a prefix tree). Details of the methods for approximate dictionary searching can be found in the surveys on this topic [11,14,19,5].

A common approach to approximate dictionary searching involves generation of a pattern full $k$-neighborhood: strings obtainable from the pattern by at most $k$ edit operations. Then, elements of the full neighborhood are searched in the dictionary for an exact match. This method is not efficient for large $k$ and/or large alphabets, because the size of the full neighborhood is $O\left(n^k|\Sigma|^k\right)$ (where $n$ and $|\Sigma|$ is the size of the pattern and the alphabet, respectively) [21].

Much shorter retrieval times can be achieved through indexing of residual strings, i.e., strings obtainable by deletions from dictionary words. Along with residual strings, it is necessary to memorize deleted characters and their positions in the original dictionary words. We call these data deletion lists. A special case of this method for $k = 1$ was described by Mor and Fraenkel in 1982 [17]. A generalization of the Mor-Fraenkel method for $k > 1$ was independently proposed by Bocek et al. [3] and Boytsov [5]. Both Bocek et al. and Boytsov suggested to index residual strings and positions of deleted characters directly (via a hash index), which is not space efficient. For $k = 1$, there are methods that have better space requirements. Mihov and Schulz [16] store one-deletion neighborhoods in the form of finite transducers. In the algorithm by Belazzougui [2], all residual words are enumerated using a minimal perfect hash function. Then, only deleted characters instead of original dictionary strings are memorized. We are not aware of any attempts to utilize compact deletion indices for $k > 1$.

An intermediate approach between the full neighborhood generation and the generation of deletion neighborhoods is a reduced alphabet neighborhood generation. In this approach, the strings over the original alphabet are mapped into strings over a smaller, i.e., reduced, alphabet. An experimental evaluation of the reduced alphabet neighborhood generation was carried out by Boytsov [5]. It is possible to combine the reduced alphabet neighborhood generation with the Mor-Fraenkel method, but we have not seen an implementation of this idea before.

All described modifications of the Mor-Fraenkel algorithm do not entail computation of the Levenshtein distance or explicit verification if the distance between the pattern and dictionary words is at most $k$. In the lossy version of the Mor-Fraenkel method descried by Karch et al. [12], deletion indices are used only as a filtering step. Instead of memorizing residual strings, deleted characters, and their positions, Karch et al. propose to keep only identifiers of original dictionary words. Consequently, at the verification step, a list of candidates should be compared directly against a search pattern through computing the Levenshtein distance. Karch et al. combine this approach with pattern partitioning: most dictionary words are divided into halves and each half is indexed separately (this approach was known already in the seventies [13,9]).

There were also attempts to blend partial neighborhood generation with tries. Cole et al. [7] introduced a $k$-errata tree, where errors are treated by recursively creating insertion, substitution, and deletion subtrees. The $k$-errata tree has a super-linear index whose size is upper bounded by $O\left(\lambda N + N \frac{(5\log_2 N)^k}{k!}\right)$, where $N$ is the number of dictionary strings. Boytsov [5] conducted an experimental evaluation of this method, which showed that the $k$-errata tree was impractical for $k > 1$.

## 1.2    Notation and Problem Formalization

We consider algorithms that operate on strings, i.e., sequences of characters over an ordered finite alphabet $\Sigma$ ($|\Sigma|$ is the size of the alphabet). The $i$-th character of the string $s$ is denoted by $s_{[i]}$. The string obtained from $s$ by deletion of the $i$-th character is denoted by $\Delta_i(s)$. A reverse operation consists in inserting character $c$ into position $i$, which introduces $c$ *before* the character $s_{[i]}$. We assume that the reader is familiar with notions of a substring as well with the concepts of a prefix, a suffix, and, a $q$-gram (a substring of the fixed length $q$).

Several implemented algorithms rely on mapping the original alphabet $\Sigma$ to a smaller alphabet $\sigma$, which is called a reduced alphabet. A projection is done using a hash function $h(c)$, which induces a character-wise projection from the set of strings over the original alphabet $\Sigma$ to the set of strings over the reduced alphabet $\sigma$ in a straightforward way.

The similarity between functions $u$ and $v$ is measured via the Levenshtein distance, which is denoted by $ED(u,v)$. It is equal to the minimum number of basic edits (insertions, deletions, and substitutions) required to convert $u$ into $v$ (and vice versa). Knowledge of algorithms to compute the Levenshtein distance is not required for understanding this paper.

Assume that $W = (s_1, s_2, \ldots, s_N)$ is an ordered set of strings, called dictionary. The search pattern and its length are denoted by $p$ and $n$, respectively. The maximum allowed edit distance is represented by $k$. The problem of approximate dictionary searching consists in retrieval of *all* dictionary strings $s_i$ such that $ED(p, s_i) \leq k$. In the *associative* version of this problem, it is necessary to find all strings $s_i$ within distance $k$ from the pattern as well as data associated with strings $s_i$ (also called satellite data). A string identifier is one well-known example of satellite data.

## 2  Method Descriptions

### 2.1  Full, Reduced, and Deletion Neighborhood

A neighborhood generation is a classic search method [10]. The full neighborhood generation entails computation of all strings within the Levenshtein distance $k$ from the search pattern $p$. These strings comprise a full $k$-neighborhood. Each element of the full $k$-neighborhood is searched for in the dictionary exactly. Because the size of the neighborhood is $O\left(n^k |\Sigma|^k\right)$ [21], this algorithm is only practical when neither of the following parameters are large: the size of the alphabet, the maximum allowed Levenshtein distance, and the pattern length.

Consider an example of the string find. The full one-neighborhood contains the following strings:

- the original string find;
- 4 strings obtained by applying a single deletion;
- $5 \times 26$ strings obtained by applying a single insertion;
- $4 \times 25$ strings obtained by applying a single substitution.

In total, the one-neighborhood contains 231 unique strings. However, the two- and the three-neighborhood of the string find contain about 20K and 1.5M unique strings, respectively.

One approach to compress the full neighborhood is to replace some characters with wildcards. Let us extend the alphabet with a wildcard pseudo-character ? that matches any alphabet character. Then, the full wildcard one-neighborhood of the word find may contain the following strings:

- the original string find;
- 4 strings obtained by a single deletion;
- 5 strings obtained by one insertion: ?find, f?ind, fi?nd, fin?d, find?;
- 4 strings obtained by one substitution: ?ind, f?nd, fi?d, fin?.

The wildcard one-neighborhood comprises only 13 strings as compared to 231 strings of the full one-neighborhood. In general, the size of the wildcard $k$-neighborhood is smaller than the size of the full $k$-neighborhood by a factor of $|\Sigma|^k$.

Another approach to compress the neighborhood is to decrease the size of the alphabet. This can be achieved through mapping of the original alphabet $\Sigma$ to a smaller (reduced) alphabet $\sigma$ via a hash function $h(c)$. Assume that $\sigma = \{0, 1\}$; $h(c)$ is equal to 0 for English letters from a to m, and is 1 for letters from n to z. Note that characters 0 and 1 can be considered as special wildcard characters that represent regular expressions [a-m] and [n-z], respectively.

In our example, the reduced string $h(\mathtt{find})$ is equal to 0010. The full neighborhood of the string 0010 has 14 unique elements, which is much smaller than the full neighborhood of the original string find. One should now be convinced that generating a wildcard/reduced alphabet neighborhood entails significant performance improvements, if we can devise algorithms to satisfy wildcard queries efficiently. In the following subsections, we discuss such algorithms.

## 2.2    A Generalization of the Mor-Fraenkel Method

In the Mor-Fraenkel method, wildcard queries are answered with a help of deletion indices. Deletion indices store deletion neighborhoods generated at index time. Consider an example, where find is a pattern string, mind is a dictionary string, and $k = 1$. Strings find and mind differ by one substitution. Furthermore, the string ?ind from the wildcard one-neighborhood of the pattern find matches the dictionary string mind.

To find all dictionary words that differ from find only in the first letter, it is sufficient to memorize all strings obtained by deletion of the first character in a special index. At search time, we simply remove the first character of the pattern string $p$ and retrieve all strings from the special index that match the shortened pattern *exactly*. In what follows, we describe a generalization of this idea. Note that Bocek et al. [3] provide an alternative description of the same approach as well as its efficient implementation (FastSS).

The indexing algorithm of the generalized Mor-Fraenkel method iterates over dictionary strings and generates their $k$-deletion neighborhoods, i.e., all strings obtainable from dictionary strings through $k$ deletions. Consider a residual string $s' = \Delta_{\tau_1}(\Delta_{\tau_2}(\dots (\Delta_{\tau_l} s) \dots)) = \Delta_{\tau_l + l - 1}(\dots (\Delta_{\tau_2 - 1}(\Delta_{\tau_1} s)) \dots)$ obtained from a dictionary string $s$ through deleting characters $s_{[\tau_1]}, s_{[\tau_2]}, \dots, s_{[\tau_l]}$ in positions $\tau_1 \leq \tau_2 \leq \dots \leq \tau_l$ ($l \leq k$). For each residual string $s'$, we memorize a triple $(s', D^s, C^s)$, where $C^s = (s_{[\tau_1]}, s_{[\tau_2]}, \dots, s_{[\tau_l]})$ stands for deleted characters and $D^s = (\tau_1, \tau_2 - 1, \dots, \tau_l - l + 1)$ represents their positions in the original string $s$. These triples, called deletion lists, are kept in an index. This index allows us to search for triples by their first elements, i.e., by residual strings.

We note the following:

- Given a triple $(s', D^s, C^s)$, the original string $s$ can be reconstructed by inserting characters $C_i^s$ at positions $D_i^s$ into the string $s'$ in the *decreasing* order of $i$.

– $D^s$ is a multiset, i.e., a set that may contain repeated elements. A multiset is characterized by its indicator function. The value of the multiset indicator function $\mathbf{1}_A(e)$ is equal to the number of times the element $e$ repeats in $A$. All multiset operations can be expressed in terms of indicator functions. In particular, the indicator of the intersection is equal to $\min(\mathbf{1}_A(e), \mathbf{1}_B(e))$ and $|A|$ (the cardinality of $A$) is equal to $\sum_{e \in A} \mathbf{1}_A(e)$.

At search time, we generate the $k$-deletion neighborhood of the pattern string $p$. Thus, we obtain pairs $(p', D^p)$, where $p'$ is a residual string obtained from $p$ by deleting characters $p_{[\rho_1]}, p_{[\rho_2]}, \dots p_{[\rho_m]}$ in positions $\rho_1 < \rho_2 < \dots < \rho_m$ and $D^p = (\rho_1, \rho_2 - 1, \dots, \rho_m - m + 1)$ is a multiset that represents $\rho_i$.

Next, we retrieve all dictionary triples $(s', D^s, C^s)$ (using the exact-search index) that satisfy the conditions:

$$p' = s'$$

$$|D^s| + |D^p| - |D^s \cap D^p| \leq k \tag{1}$$

Finally, dictionary strings are reconstructed from triples satisfying Condition (1).

## 2.3    A Compact Version of the Mor-Fraenkel Method

Explicit indexing of triples $(s', D^s, C^s)$ – defined in Subsection 2.2 – requires a lot of RAM. A more space efficient version was proposed by Belazzougui for the case $k = 1$. He suggested to enumerate all residual strings $s'$ using a minimal perfect hash function [2]. The minimal perfect hash function $f(s)$ maps $m$ strings to integer values from 1 to $m$ without collisions. During indexing, we convert triples $(s', D^s, C^s)$ into triples $(f(s'), D^s, C^s)$ and index the latter using first elements (i.e., values of the perfect hash function) as keys.

The retrieval algorithm is almost identical to that described in Subsection 2.2. At search time, we compute all pairs $(p', D^p)$, where $p'$ is a residual string obtained from $p$ by deleting up to $k$ characters. Positions of deleted characters are defined by the multiset $D^p$. Then, we retrieve all dictionary triples $(f(s'), D^s, C^s)$ such that:

$$f(s') = f(p')$$

$$|D^s| + |D^p| - |D^s \cap D^p| \leq k. \tag{2}$$

Note that the first element of the triple is the hash value of the unknown string $s'$. If $s' = p'$, the triple represents a dictionary string $s$ such that $ED(p, s) \leq k$. In this case, $s$ can be obtained by inserting $C_i^s$ into $p'$ at positions $D_i^s$ (in the decreasing order of $i$). However, if $s' \neq p'$, the triple represents a false positive.

Two cases are to be considered. In the first case, the residual pattern string belongs to the set of residual dictionary strings computed during indexing. By the definition of the perfect hash function, $f(s') = f(p')$ implies $s' = p'$. Thus, the retrieved triple represents the dictionary string $s$ such that $ED(p, s) \leq k$. The string $s$ can be recovered from the residual pattern string $p'$, the multiset $D^s$, and the vector $C^s$ as described previously.

In the second case, $p'$ is not a residual dictionary string. Thus, $p' \neq s'$. This case signifies a false positive. As noted by Belazzougui, it can be detected by constructing the string $s''$ from $\{p', D^s, C^s\}$ as described previously and checking whether the constructed string belongs to the dictionary.

If $s''$ is a dictionary string, then all the triples satisfying Condition (2) define dictionary strings $s$ such that $ED(p,s) \leq k$. If the constructed string does not belong to the dictionary, none of the triples satisfying Condition (2) represent a dictionary string $s$ such that $ED(p,s) \leq k$. Thus, checking whether a reconstructed string $s''$ belongs to the dictionary has to be done only once for every residual pattern string $p'$.

To obtain a compact index of deletion lists, Belazzougui recommends to store triples in the increasing order of hash values $f(s')$. For each hash value $i$, the offset of the first triple $(f(s'), D^s, C^s)$ such that $f(s') = i$ is stored in the offset table $T(i)$. Because the offsets in $T(i)$ is a sequence of non-decreasing integer values, one can efficiently compress $T(i)$. An experimental survey of methods for compact representation of directly addressable ordered sets is given by Brisaboa et al. [6]. In our work, we rely on a simple folklore sampling method, which allows us to compress the offset table $T(i)$ to about 30-50% of its original size.

To conclude this subsection, we note that in our implementation the perfect hash functions are computed using the CMPH library [4].[1] For our data, CMPH fails to generate a perfect hash function when the number of residual strings is large (approximately 100M). To overcome this difficulty, we employ a two-level scheme, where residual strings are divided into shards using a regular hash function. Then, we create a perfect hash function separately for each shard. It is noteworthy, that this approach allows one to construct a perfect hash function for arbitrarily large sets of strings. In addition, dividing the index into shards simplifies updates.

## 2.4   Reduced Alphabet Neighborhood Generation

The indexing algorithm of the reduced alphabet neighborhood generation employs a hash function $h(c)$ to convert original dictionary strings $s_i$ into their projections $h(s_i)$, which are strings in the reduced alphabet. Then, dictionary strings are organized into buckets based on the values of $h(s_i)$ so that each bucket contains strings with same values of $h(s_i)$. This allows us to efficiently retrieve original strings $s_i$ using their projections $h(s_i)$ as search patterns.

At search time, the pattern $p$ is converted into $r = h(p)$. Then, we create a full $k$-neighborhood of the reduced pattern $r$ (using characters from the reduced alphabet $\sigma$). All dictionary strings $s$ such that $ED(p,s) \leq k$ are contained in buckets corresponding to strings from the generated neighborhood. This step provides a list of candidate strings. Because the size of the reduced alphabet is (much) smaller than that of the original alphabet, computation of the reduced alphabet requires little time.

---

[1] It can be downloaded from http://cmph.sourceforge.net/

In the second step, the candidate strings are compared with the original pattern. A naive implementation of the verification step involves computation of the Levenshtein distance. A more efficient approach is to generate an additional wildcard neighborhood, i.e., the wildcard neighborhood of the original pattern.

The generation of two neighborhoods is synchronized in the following manner:

- If we substitute the $i$-th character of the reduced pattern $r$, we replace the $i$-th character of the original pattern $p$ with the wildcard symbol ?;
- Similarly, if we insert a character at position $i$ of $r$, we insert the wildcard ? at position $i$ of $p$;
- If we delete the $i$-th character of $r$, we also delete the $i$-th character of the original pattern $p$.

Note that this procedure generates pairs of patterns that have same lengths.

Consider the binary reduced alphabet and the hash function $h(c)$ defined in Subsection 2.1. Assume that the pattern $p = $ ind is a misspelled version of the dictionary string find. Then, $r = h(\text{ind}) = 010$ and $h(\text{find}) = 0010$. The reduced-alphabet one-neighborhood of $r$ contains the string 0010, which is obtained by inserting 0 into the first position of the reduced pattern. The respective element from the "parallel" neighborhood is equal to ?ind. We use 0010 to identify a bucket that contains the string find. Then, the element ?ind of the second neighborhood is used to compare ind with find. For this purpose, we treat ?ind as a simple regular expression where ? matches any alphabet character. Whenever a dictionary string in the bucket matches $p$ within $k = 1$ errors, it should match such regular expression exactly. This match can be verified efficiently (in time proportional to the length of $s$) without computing the Levenshtein distance.

## 2.5   A Hybrid of the Mor-Fraenkel Method and the Reduced Alphabet Neighborhood Generation

The Mor-Fraenkel method can be blended with the reduced alphabet neighborhood generation. The indexing process of this hybrid algorithm starts with creating a reduced alphabet index outlined in Subsection 2.4: The dictionary strings $s_i$ are divided into buckets based on their projections $h(s_i)$ (to the set of reduced alphabet strings). Projections $h(s_i)$ are stored in the form of a dictionary. In the second indexing step, this dictionary is indexed using a compact version of the Mor-Fraenkel method (see Subsection 2.3). One advantage of this approach is that triples $(f(s'), D^s, C^s)$ can be readily compressed. For example, in our experiments we use $|\sigma| = 8$. Thus, each element of $C^s$ can be encoded with 3 bits.

The search algorithm is divided into two steps and involves the parallel generation of two wildcard neighborhoods. This first step of the search algorithm is a modification of the Mor-Fraenkel search method. As described in Subsection 2.2, we create residual patterns $\{p'\}$ by applying up to $k$ deletions. Positions of deleted characters associated with $\{p'\}$ are defined by multisets $\{D^p\}$. In addition, we create residual patterns $\{r'\}$ by deleting characters from the reduced pattern $r = h(p)$ in the same positions as in $\{p\}$.

For each $r'$, we retrieve memorized triples $(f(s'), D^s, C^s)$ such that $f(r') = f(s')$ and $|D^s| + |D^{r'}| - |D^s \cap D^{r'}| \leq k$ ($s'$ is a string in the reduced alphabet). Then, we construct a string $s''$ by inserting characters $C^s$ into the residual string $r'$ in positions $D_i^s$ in the decreasing order of $i$. The result is a string $r''$. We also modify the residual string $p'$, obtained from the non-reduced pattern $p$. To this end, wildcard character ? is inserted into positions $D_i^s$ in the decreasing order of $i$. The result is the pattern $p''$ that contains zero or more wildcards. The string $r''$ defines a bucket with candidate strings, which are exhaustively compared with the simple regular expression defined by $p''$.

## 2.6   Associativity Consideration

It can be seen that methods defined in Subsections 2.4-2.5 can handle associated data by simply storing it in the buckets (or pointers thereto). Because both methods involve full scanning of the buckets with candidate strings, retrieval of associated data does not have a performance penalty.

However, the compact version of the Mor-Fraenkel method that uses perfect hashing is capable of retrieving only strings themselves. Retrieval of associated data can be supported in two ways. In a more space efficient approach, associated data (or pointers thereto) are stored in the dictionary. Then, for every string generated during the verification step, we have to search the dictionary for an exact match, even though we already know that the generated string must belong to the dictionary. According to our experiments, these additional lookups almost double retrieval time. In a second approach, compressed string identifiers are kept in the deletion lists. This requires an approximately two times larger index, but retrieval time will remain the same.

## 3   Experiments

### 3.1   Experimental Setup

Experiments are carried out on a laptop with a 2 Ghz Dual Core Intel Processor, 3 Gb of RAM, and 1 Mb of L2 cache. This laptop is running a 32-bit Linux (kernel version 2.6.x).

We use some of the data sets published by Boytsov [5]: synthetic English dictionaries, frequent words from the ClueWeb09 collection, and DNA sequences extracted from the human genome (of length 11). Each type of the data set has 5 dictionaries with 0.2M, 0.4M, 0.8M, 1.6M, and 3.2M strings. We created new sets of search patterns containing up to 10K, by applying up to $k$ random edits to dictionary strings. These larger test sets allow us to reduce sampling uncertainty in calculating average retrieval times.

All search methods are implemented in C/C++ with correctness of implementation verified by black-box testing. We index a small dictionary and generate a set of strings by applying $i \leq k$ random edits to dictionary words. Then, we check if the algorithm is capable of finding original dictionary words using modified strings as search patterns.

In our experiments, we determine how the following performance character-istics depend on $k$: the average in-memory retrieval time and the index size. To estimate index memory requirements, we measure the amount of space occupied by a serialized version of the index. For FastSS this method produces a biased es-timate: we correct it through multiplying by 2, which is a coefficient empirically determined using the Unix utility `top`.

## 3.2   Evaluated Methods

We compared the performance of super-linear indices with several methods, in particular with the fastest methods evaluated by Boytsov [5]. We benchmarked the following algorithms:[2]

- FastSS [3], which is straightforward generalization of the Mor-Fraenkel method (see Subsection 2.2). Deletion lists are not compressed.
- A new implementation of the Mor-Fraenkel method with compact indices (see Subsection 2.3). It employs the perfect hash library CMPH [4].[3] We compress deletion lists as follows: for natural language data, each deleted character as well as its position is encoded using 6 bits. For DNA-data, a deleted character together with its position occupy 8 bit.
- The reduced alphabet neighborhood generation implemented by Boytsov [5] (see Subsection 2.4). In the case of ClueWeb09 data, we use $|\Sigma| = 5$ and $|\Sigma| = 3$, otherwise. This method works well only for large and medium alphabets and is not used for the DNA data set.
- The full neighborhood generation (see Subsection 2.1), which is used only for the DNA data set.
- The hybrid of the reduced alphabet neighborhood generation and the Mor-Fraenkel method with $|\Sigma| = 8$ (see Subsection 2.5). Characters and their positions in deletion lists are encoded using 3 and 5 bits, respectively. For a few patterns longer than $31 - k$ character, we resort to the reduced alphabet neighborhood generation.
- The FB-trie proposed by Mihov and Schulz [16] (Boytsov's implementation [5]). It employs a pair of tries: a regular one and a trie built over reversed strings. At search time, the method looks for the original pattern in the reg-ular trie, and for the reversed pattern in the trie built over reversed strings. A first part of either the original or the reversed pattern should match a trie prefix with at most $\lfloor k/2 \rfloor < k$ errors. This is a well-known pattern partitioning approach [13,9].
- Two modifications of $q$-gram methods. One is implemented by Boytsov [5] and another is provided with the Flamingo package written by Behm et al. [1]. For brevity, we report only the best time achieved by one of the $q$-gram methods.

---

[2] The source files are available at: `http://boytsov.info/src/`.
[3] `http://cmph.sourceforge.net/`

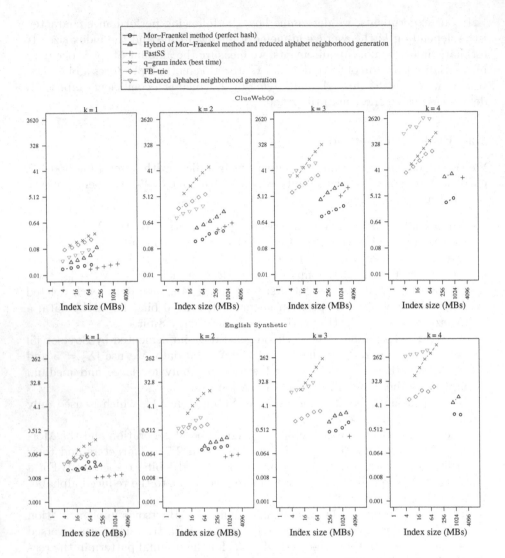

**Fig. 1.** Relationship between the average retrieval time and the index size (log-scale on both axes). Each series of connected dots represents results for dictionaries of at most five sizes: 0.2M, 0.4M, 0.8M, 1.6M, and 3.2M (from left to right).

## 3.3 Experimental Results

Figures 1 and 2 shows the relationship between the average retrieval time and the index size. Each series of connected dots represents results for a set of dictionaries of increasing size. In most cases, the dots from left to right correspond to the dictionaries of five sizes: 0.2M, 0.4M, 0.8M, 1.6M, and 3.2M. Some connected series contain fewer dots, because larger indices do not fit into RAM. Note that in the case of associative searching, either the average retrieval time or the index size of the Mor-Fraenkel method based on perfect hashing would be twice of that presented in Figures 1-2. The other methods can support associative searching without a penalty in performance or memory requirements (see Subsection 2.6).

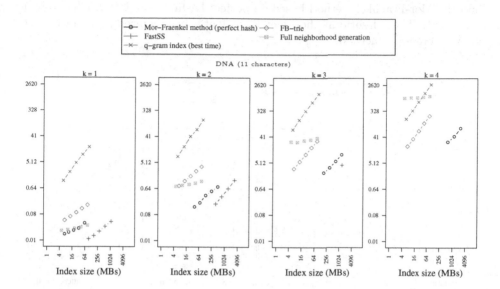

**Fig. 2.** Relationship between the average retrieval time and the index size (log-scale on both axes). Each series of connected dots represents results for dictionaries of at most five sizes: 0.2M, 0.4M, 0.8M, 1.6M, and 3.2M (from left to right).

One can immediately see that all three modifications of the Mor-Fraenkel method (which includes FastSS) are more efficient than other methods in almost all cases. In particular, they are:

— about two orders of magnitude faster than $q$-gram based methods, which are often considered as good benchmarks;
— up to an order of magnitude faster than the FB-trie.

This efficiency comes at the price of huge indices and long indexing times (up to one hour for the variant based on perfect hashing). Consider the panel in Figure 1 corresponding to the case of $k = 1$. The second dot in the FastSS series represents the index for the second largest dictionary (0.4M strings). It has the size 100MB, which is larger than a $q$-gram index built for the dictionary with 3.2M strings. However, the index size of the Mor-Fraenkel method based on perfect hashing is only 8MB, or about 1/10 of the FastSS index. Taking into account that about 25% reduction is achieved through lightweight compression of deletion lists (both characters and positions occupy 6 bits each), we obtain that the use of perfect hashing alone lead to about 8-fold reduction in index sizes as compared to straightforward memorization of deletion neighborhoods. For larger $k$ the difference is approximately 4-fold.

One can also see that the hybrid of the Mor-Fraenkel method and the reduced alphabet neighborhood generation is not a very practical method. Even though the hybrid method significantly improves over the reduced alphabet neighborhood generation (especially for larger $k$), it is up to an order of magnitude slower than the Mor-Fraenkel method based on perfect hashing (see $k = 4$, ClueWeb09 data). In that, the hybrid method has equivalent space requirements to those of the Mor-Fraenkel method.

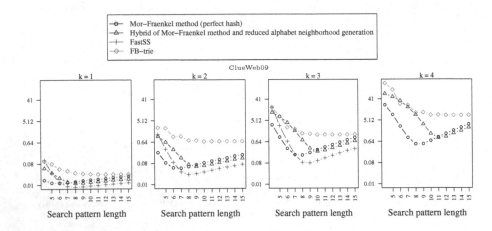

**Fig. 3.** Relationship between the average retrieval time and pattern length (log-scale on time axis)

Consider the case of DNA data. For small dictionaries, the variants of Mor-Fraenkel methods are among the fastest algorithms. As the number of dictionary strings grows, performance of these methods deteriorates. For $k \leq 2$, it becomes equivalent to that of full neighborhood generation. We believe that this fact can be explained by a density effect (see Section C.3.3 in the paper by Boytsov [5]). In our case, the number of unique 11-character DNA sequences is about 4M. The largest dictionary with 3.2M entries contains most of them and, thus, is very dense. Consequently, the algorithm has a low filtering efficiency. Note that for $k \geq 3$ and DNA data, the Mor-Fraenkel method outperforms the full neighborhood generation, but it has the equivalent performance to that of the FB-trie.

We conducted an additional experiment to study the relationship between the average retrieval time and the pattern length. To this end, we use the smallest ClueWeb09 dictionary (0.2M strings) and patterns with length from 4 to 15. According to Figure 3, the average retrieval time of all Mor-Fraenkel methods first decreases until $n \approx 8$. Afterwards, it increases monotonically. The FB-trie utilizes pattern partitioning, which is essentially filtering by word halves. The longer is the pattern, the better is filtering efficiency. Consequently, the average retrieval time of the FB-trie decreases monotonically with $n$. For long patterns, $k = 1$, and $k = 3$, performance of Mor-Fraenkel methods is equivalent to that of the FB-trie. Thus, Mor-Fraenkel methods would be most useful for short and medium-size patterns.

## 4    Conclusions

Mor-Fraenkel methods have tremendous space requirements: The index size grows exponentially with $k$. This also applies to FastSS, which belongs to the family of Mor-Fraenkel methods. We have empirically confirmed that space requirements can be 4-8 times lower if perfect hashing is employed (the idea proposed by Belazzougui [2]). Given that typical servers are now equipped with 8-32 Gb of memory, this method is applicable to natural language dictionaries containing several million entries. At the same time, the efficiency of the Mor-Fraenkel method based on perfect hashing is similar to that of the straightforward implementation of the Mor-Fraenkel method, which indexes deletion neighborhoods directly. Both the straightforward and perfect-hash implementations outperform our fastest benchmarks in most cases.

Mor-Fraenkel methods work best for small and medium patterns (at most 10 characters). For longer search strings one should employ a pattern partitioning strategy similar to the one used by Karch et al. [12]. However, it remains to be determined which pattern partitioning strategy would be most efficient. Another open question is whether (and to what extent) one can improve performance of a trie-based method through precomputing wildcard neighborhoods at index time. One such algorithm was proposed by Cole et al. [7], but we are unaware of any space efficient implementation of this (or similar) method.

**Acknowledgments.** I am very grateful to my wife Anna for editorial assistance.

## References

1. Behm, A., Vernica, R., Alsubaiee, S., Ji, S., Lu, J., Jin, L., Lu, Y., Li, C.: UCI Flamingo Package 4.0 (2010)
2. Belazzougui, D.: Faster and Space-Optimal Edit Distance "1" Dictionary. In: Kucherov, G., Ukkonen, E. (eds.) CPM 2009. LNCS, vol. 5577, pp. 154–167. Springer, Heidelberg (2009)
3. Bocek, T., Hunt, E., Stiller, B.: Fast similarity search in large dictionaries, Technical report No. ifi-2007.02, Department of Informatics (IFI), University of Zurich (2007)
4. Botelho, F.C.: Near-Optimal Space Perfect Hashing Algorithms. PhD thesis, Graduate Program in Computer Science, Federal University of Minas Gerais, Brazil (2008)
5. Boytsov, L.: Indexing methods for approximate dictionary searching: Comparative analysis. J. Exp. Algorithmics 16, 1.1:1.1–1.1:1.91 (2011)
6. Brisaboa, N.R., Ladra, S., Navarro, G.: Directly Addressable Variable-Length Codes. In: Karlgren, J., Tarhio, J., Hyyrö, H. (eds.) SPIRE 2009. LNCS, vol. 5721, pp. 122–130. Springer, Heidelberg (2009)

7. Cole, R., Gottlieb, L.A., Lewenstein, M.: Dictionary matching and indexing with errors and don't cares. In: STOC 2004: Proceedings of the Thirty-Sixth Annual ACM Symposium on Theory of Computing, pp. 91–100. ACM (2004)

8. Damerau, F.: A technique for computer detection and correction of spelling errors. Communications of the ACM 7(3), 171–176 (1964)

9. Doster, W.: Contextual postprocessing system for cooperation with a multiple-choice character-recognition system. IEEE Trans. Comput. 26, 1090–1101 (1977)

10. Gorin, R.E.: SPELL: Spelling check and correction program, Online documentation: Describes operation of PDP-10 SPELL program (1971), `http://pdp-10.trailing-edge.com/decuslib10-03/01/43,50270/spell.doc.html` (accessed May 28, 2012)

11. Hall, P., Dowling, G.: Approximate string matching. ACM Computing Surveys 12(4), 381–402 (1980)

12. Karch, D., Luxen, D., Sanders, P.: Improved fast similarity search in dictionaries. CoRR abs/1008.1191 (2010)

13. Knuth, D.: The Art of Computer Programming. Sorting and Searching., 1st edn., vol. 3. Addison-Wesley (1973)

14. Kukich, K.: Technique for automatically correcting words in text. ACM Computing Surveys 24(2), 377–439 (1992)

15. Levenshtein, V.: Binary codes capable of correcting deletions, insertions, and reversals. Doklady Akademii Nauk SSSR 163(4), 845–848 (1965)

16. Mihov, S., Schulz, K.U.: Fast approximate string search in large dictionaries. Computational Linguistics 30(4), 451–477 (2004)

17. Mor, M., Fraenkel, A.S.: A hash code method for detecting and correcting spelling errors. Communications of the ACM 25(12), 935–938 (1982)

18. Navarro, G.: A guided tour to approximate string matching. ACM Computing Surveys 33(1), 31–88 (2001)

19. Owolabi, O.: Dictionary organizations for efficient similarity retrieval. Journal of Systems and Software 34(2), 127–132 (1996)

20. Sankoff, D.: The early introduction of dynamic programming into computational biology. Bioinformatics 16(1), 41–47 (2000)

21. Ukkonen, E.: Approximate String Matching Over Suffix Trees. In: Apostolico, A., Crochemore, M., Galil, Z., Manber, U. (eds.) CPM 1993. LNCS, vol. 684, pp. 228–242. Springer, Heidelberg (1993)

# Visual Image Search: Feature Signatures or/and Global Descriptors

Jakub Lokoč[1], David Novák[2], Michal Batko[2], and Tomáš Skopal[1]

[1] SIRET Research Group, Faculty of Mathematics and Physics,
Charles University in Prague
{lokoc,skopal}@ksi.mff.cuni.cz
[2] Faculty of Informatics, Masaryk University in Brno
{david.novak,batko}@fi.muni.cz

**Abstract.** The success of content-based retrieval systems stands or falls with the quality of the utilized similarity model. In the case of having no additional keywords or annotations provided with the multimedia data, the hard task is to guarantee the highest possible retrieval precision using only content-based retrieval techniques. In this paper we push the visual image search a step further by testing effective combination of two orthogonal approaches – the MPEG-7 global visual descriptors and the feature signatures equipped by the Signature Quadratic Form Distance. We investigate various ways of descriptor combinations and evaluate the overall effectiveness of the search on three different image collections. Moreover, we introduce a new image collection, TWIC, designed as a larger realistic image collection providing ground truth. In all the experiments, the combination of descriptors proved its superior performance on all tested collections. Furthermore, we propose a re-ranking variant guaranteeing efficient yet effective image retrieval.

## 1 Introduction

With the increasing volumes of multimedia data available over the internet, the Content-based Image Retrieval Systems (CBIR) [10,11] steadily become more and more important. Even though for some of the data an annotation is available, the content-based paradigm (possibly combined with the keyword search) might provide more precise retrieval than the keyword search alone. This fact was recently confirmed by Google that added content-based image search to the classic keyword image search engine. However, unlike keyword search, the technology of content-based image retrieval is extremely diverse as there have been thousands models proposed and even more studies performed [10]. The differences reflect various demands, such as general image-matching technologies and applications vs. very specialized systems tuned for a specific application, use various feature extraction types designed for measuring global or local similarity, and so on. The image descriptors span from image fingerprints (hashes) for near-duplicate search, over local image features, to descriptors of global features.

G. Navarro and V. Pestov (Eds.): SISAP 2012, LNCS 7404, pp. 177–191, 2012.
© Springer-Verlag Berlin Heidelberg 2012

In this paper, we deal with MPEG-7 global visual descriptors and so-called image feature signatures. The MPEG-7 visual descriptors use standardized description of image content that proved to provide good retrieval effectiveness in image retrieval applications. Their main property is they describe global image features, such as color, texture or shape distribution, among others. On the other hand, the recently proposed feature signatures allow to aggregate local features into a compact form. It has been shown that feature signatures provide more flexible similarity search than MPEG-7 descriptors, while they offer less complex matching than local features developed for image classification, e.g., SIFTs.

**Paper Contribution**
Both MPEG-7 descriptors and feature signatures have their strong and weak points. In this paper we compare and synergistically combine MPEG-7 descriptors with image feature signatures in order to reach ultimate effectiveness. In particular, we

- evaluate on three different image collections the effectiveness of standard global visual descriptors (and their combinations) and complex feature signatures used with the signature quadratic form distance (SQFD),
- combine the MPEG-7 global descriptors with the feature signatures in various ways which improves the overall effectiveness of the search on all tested collections,
- employ the re-ranking concept, such that the MPEG-7 descriptors are used for "cheap" pre-selection of image candidates and then the (rather small) result is re-ranked using the time-consuming SQFD based on feature signatures, resulting thus in a very efficient search mechanism,
- and introduce a new image collection for CBIR effectiveness evaluation.

## 2   Preliminaries and Related Work

When searching multimedia databases in a content-based way, users issue similarity queries by selecting multimedia objects or by sketching the intended object contents. Given an example multimedia object or sketch $q$, the multimedia database $\mathbb{S} \subset \mathbb{U}$ (where $\mathbb{U}$ is the object universe) is searched for the most related objects with respect to the query by measuring the similarity between the query and each database object by means of a distance function $\delta$. As a result, the multimedia objects with the lowest distance to the query are returned to the user. In particular, a *range query* $(q, r)$, $q \in \mathbb{U}$, $r \in \mathbb{R}^+$, reports all objects in $\mathbb{S}$ that are within a distance $r$ to $q$, that is, $(q, r) = \{x \in \mathbb{S} \mid \delta(x, q) \le r\}$. The subspace defined by $q$ and $r$ is called the *query ball*. Another popular similarity query is the *k nearest neighbors query* ($k$-NN(q)). It reports the $k$ objects from $\mathbb{S}$ closest to $q$. That is, it returns the set $\mathbb{C} \subseteq \mathbb{S}$ such that $|\mathbb{C}| = k$ and $\forall x \in \mathbb{C}, y \in \mathbb{S} - \mathbb{C}, \delta(x, q) \le \delta(y, q)$. The $k$-NN query also defines a query ball $(q, r)$, but the distance $r$ to the $k^{th}$ NN is not known beforehand. In the following paragraphs, we describe two different model representations used in this paper.

## 2.1 MPEG-7 Global Visual Descriptors

The global visual descriptors are the fundamental instruments to measure the overall similarity of the digital images' content. In this work, we use five well-established descriptors from the MPEG-7 standard [24] that capture various image characteristics. There is a function defined for each of the descriptors [21] to measure the *distance* (dissimilarity) $\delta$ between two instances of that descriptor.

**Scalable Color** is derived from a color histogram in the Hue-Saturation-Value color space with fixed space quantization. We used the 64 coefficients version of this descriptor. The distance between two scalable color instances is measured by the $L_1$ metric (sum of absolute differences).

**Color Structure** aims at identifying localized color distributions using a $8 \times 8$ pixels structuring matrix that slides over the image. This descriptor can distinguish between two images having similar amount of pixels of a specific color, if structures of these pixels differ in these images. The $L_1$ metric is used to compute descriptors distances.

**Color Layout** descriptor is obtained by applying the Discrete cosine transform on a 2-D array (usually $8 \times 8$ blocks) of local representative colors in three color channels (Y, Cb, and Cr). The distance between two objects is computed as a sum of $L_2$ distances in each of the three color space components.

**Edge Histogram** represents the local-edge distribution in the image. The image is subdivided into $4 \times 4$ sub-images and edges in each sub-image are categorized into five types: vertical, horizontal, $45°$ diagonal, $135°$ diagonal, and non-directional edges. This results in 80 coefficients (5 values for each of the 16 sub-images) representing the local edge histograms. Further, the semi-global and the global histograms can be computed based on the local histogram and the distance is computed as a sum of weighted sub-sums of absolute differences for the local, semi-global and global histograms.

**Region Shape** descriptor considers the whole region of the shapes on the image. The descriptor works by "decomposing" the shape into a number of orthogonal 2-D basis functions defined by the Angular Radial Transformation (ART) [24]. The descriptor is a vector of normalized magnitudes of the ART coefficients and the distance is calculated using the $L_1$ norm.

## 2.2 Descriptors Consisting of Local Features

The conventional feature descriptors, such as MPEG-7 visual descriptors, aggregate and store these properties in *feature histograms*, which can be compared by vectorial distances [17,26]. The problem is, that for both simple and complex images there is the same number of bins, which does not reflect the complexity of the images. From this point of view, the *feature signatures* are more flexible choice to describe the image content.

**Feature Signatures.** Unlike conventional feature histograms, feature signatures are frequently obtained by clustering the objects' properties, such as color, position, texture, or other more complex features [12,23], within some feature space and storing the cluster representatives and weights. Thus, given a feature space $\mathbb{F}$, the *feature signature $S^o$* of a multimedia object $o$ is defined as a set of tuples from $\mathbb{F} \times \mathbb{R}^+$ consisting of representatives $r^o \in \mathbb{F}$ and weights $w^o \in \mathbb{R}^+$.

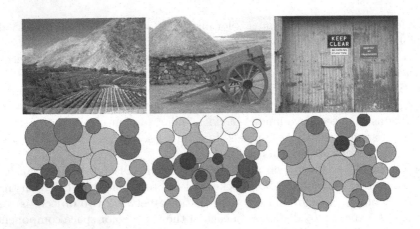

**Fig. 1.** Three example images with their corresponding feature signature visualizations

We depict an example of image feature signatures according to a feature space comprising position, color and texture information, i.e. $\mathbb{F} \subseteq \mathbb{R}^7$, in Figure 1. For this purpose, we applied a $k$-means clustering algorithm where each representative $r_i^o \in \mathbb{F}$ corresponds to the centroid of the cluster $\mathcal{C}_i^o \subseteq \mathbb{F}$, i.e., $r_i^o = \frac{\sum_{f \in \mathcal{C}_i^o} f}{|\mathcal{C}_i^o|}$, with relative frequency $w_i^o = \frac{|\mathcal{C}_i^o|}{\sum_i |\mathcal{C}_i^o|}$. We depict the feature signatures' representatives by circles in the corresponding color. The weights are reflected by the diameter of the circles. As can be seen in this example, feature signatures adjust to individual image contents by aggregating the features according to their appearance in the underlying feature space.

**Signature Quadratic Form Distance.** The Signature Quadratic Form Distance (SQFD) [6] is an adaptive distance-based similarity measure, generalizing the classic vectorial Quadratic Form Distance (QFD) [16] for feature signatures. It is defined as follows.

**Definition 1 (SQFD).** *Given two feature signatures $S^q = \{\langle r_i^q, w_i^q \rangle\}_{i=1}^n$ and $S^o = \{\langle r_i^o, w_i^o \rangle\}_{i=1}^m$ and a similarity function $f_s : \mathbb{F} \times \mathbb{F} \to \mathbb{R}$ over a feature space $\mathbb{F}$, the signature quadratic form distance $\mathrm{SQFD}_{f_s}$ between $S^q$ and $S^o$ is defined as:*

$$\mathrm{SQFD}_{f_s}(S^q, S^o) = \sqrt{(w_q \mid -w_o) \cdot A_{f_s} \cdot (w_q \mid -w_o)^T},$$

where $A_{f_s} \in \mathbb{R}^{(n+m) \times (n+m)}$ is the similarity matrix arising from applying the similarity function $f_s$ to the corresponding feature representatives, i.e., $a_{ij} = f_s(r_i, r_j)$. Furthermore, $w_q = (w_1^q, \ldots, w_n^q)$ and $w_o = (w_1^o, \ldots, w_m^o)$ form weight vectors, and $(w_q \mid -w_o) = (w_1^q, \ldots, w_n^q, -w_1^o, \ldots, -w_m^o)$ denotes the concatenation of weights $w_q$ and $-w_o$.

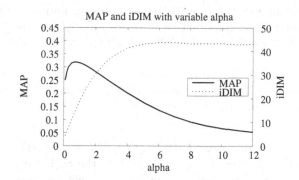

**Fig. 2.** The impact of $\alpha$ on the mean average precision (MAP) and intrinsic dimensionality (iDIM)

The similarity function $f_s$ is used to determine similarity values between all pairs of representatives from the feature signatures. In our implementation, we use the similarity function $f_s(r_i, r_j) = e^{-\alpha L_2(r_i, r_j)^2}$, where $\alpha$ is a constant for controlling the precision-indexability tradeoff, as investigated in previous works [4,19], and $L_2$ denotes the Euclidean distance. In particular, the lower values of $\alpha$ lead to better indexability (allowing fast search), that is, to lower values of so-called *intrinsic dimensionality* (iDIM) [8]. However, with lower values of the $\alpha$ parameter also the *mean average precision* (MAP) decreases, see an example for TWIC database in Figure 2. On the contrary, the best mean average precision values can be reached for already high $\alpha$ (e.g., $\alpha > 0.5$ in the figure), where the SQFD space is no longer indexable. In such cases the parallel implementation could be the only feasible way to significantly speedup the search, especially when GPU processing is employed [18]. In the following section we briefly summarize the indexing methods used for efficient retrieval.

## 2.3   Efficiency of the Retrieval

In this section, we briefly analyze how demanding is indexing and searching of the above mentioned descriptors, while we mention ways of how to speed up the search. Details can be found in the referenced literature.

**MPEG-7 Visual Descriptors.** The described MPEG-7 visual descriptors are a standard means for measuring global visual similarity of images. Indexing and searching of such data can be relatively efficient because the representations of the features is not very space demanding (all five mentioned descriptors together require about 1 kB of memory) and all the respective distance functions are based on $L_p$ metrics. Individual descriptors can be combined by a (weighted) sum of their respective distances and the result remains a metric space. On average, the time of a single distance computation of this descriptor combination is about 0.01 ms.

Number of recent works on metric-based indexing and searching presented efficiency experiments on combination of several global features [2,14,25]. The results indicate that evaluation of $k$-NN on such spaces is relatively efficient for both precise and approximate similarity search: over 50 % of the indexed data can be pruned by sophisticated precise metric access methods and about 90 % of the precise $k$-NN answer can be obtained by accessing 5–10 % of the indexed data by sophisticated approximate techniques.

**Indexing Feature Signatures with SQFD.** When processing content-based similarity queries by the naïve sequential search, the SQFD distance has to be evaluated for each database object individually. Unlike the cheap $L_p$ distances, the SQFD is of more than quadratic time complexity, so the sequential search, sometimes acceptable for $L_p$ distances, is impractical for SQFD even on a moderately sized database. Although it has been shown that the SQFD is a generalization [7] of the well-known Quadratic Form Distance [16], recent approaches indexing the data by a homeomorphic mapping into the Euclidean space [27] cannot be applied to the SQFD, as the similarity matrix changes from computation to computation.

Nevertheless, recent papers showed that SQFD can be indexed by metric access methods [4] and ptolemaic indexing [19], achieving a speed-up of up to two orders of magnitude with respect to the sequential scan.

### 2.4  Descriptor Effectiveness: Related Work

There are a number of works studying the effectiveness of global MPEG-7 descriptors and their combination for both general visual image search [13,29,3,1] and more specific application [9,28]. We can draw the following general conclusion from these studies: the descriptors can well serve for a relatively fast global visual image search and they were successfully applied in a number of applications. Specific selection or combination of descriptors depends on specific characteristics of the dataset.

Visual feature signatures, especially in combination with the SQFD measure, draw attention recently and effectiveness of this approach was tested on several CBIR collections [7,5]. These works focus on comparison of various similarity measures for this type of descriptors and they identified SQFD as superior. To the best of our knowledge, there exists no work that would compare effectiveness of feature signatures (with SQFD) and standard MPEG-7 descriptors for global visual search or that would combine these two approaches.

**Fig. 3.** The datasets used in the experiments, each dataset represented by two classes/topics

# 3    Global Descriptors and Feature Signatures

The following section is the key part of the paper. First, we describe the experiment settings, i.e., datasets, used descriptors and employed evaluation metrics. Then we evaluate the effectiveness of the individual approaches, plus we evaluate and describe various combinations of global descriptors and feature signatures.

## 3.1    Experiment Settings

**Datasets.** To conduct the experiments, first the ALOI dataset [15] comprising 72,000 images and the Corel Wang dataset [30] comprising 1,000 images were considered. Both datasets provide the ground truth in the form of classes containing particular images. The ALOI dataset consists of 1,000 classes where each class represents one object captured under various viewing angles. Six example images representing two classes in the ALOI dataset are depicted in the first column of Figure 3. Since all the images have black background which reduces the noise information and the classes are very homogeneous, the similarity search task in ALOI dataset is quite simple. The Corel Wang dataset consists of more heterogeneous images selected from ten different topics (see the second column of Figure 3). Such dataset can verify the proposed methods more thoroughly.

However, the Corel Wang dataset is quite small and not all images in particular topics are visually similar. Therefore, we have decided to create and introduce a new dataset called *Thematic Web Images Collection* (TWIC) comprising 11,555 images divided among 200 classes [20]. The TWIC dataset is intended as an alternative to ALOI – each class consists of visually similar objects but the background is heterogeneous. Six images representing two classes in the TWIC dataset are depicted in the third column of Figure 3. We may observe that in one class there is one central object on various backgrounds, which more corresponds to real requirements of the visual similarity search tasks.

To create the TWIC dataset, we have first selected several domains (e.g., Buildings, Flags, Mammals, Ocean, etc.) and for each domain we have selected around fifty keywords from that domain. Having these several hundreds of keywords where each keyword represents one image class, we have started to query the google images engine. Such keywords that created a homogeneous google image search result[1] were saved, i.e., the keyword and first two-hundred links to the corresponding images. Then, we have manually filtered all images that were not visually coherent from each image class and selected only the classes containing more than fifty objects obtaining finally 11,555 images divided among 200 classes (keywords). For more details see [20].

**Descriptors.** In the experiments, we used the five MPEG-7 descriptors described in Section 2.1 [24] together with the recommended distance measures. Standard XM library was used for extraction [21].

To create feature signatures, we have extracted seven-dimensional features $(L, a, b, x, y, \chi, \epsilon) \in \mathbb{F}$ including color $(L, a, b)$, position $(x, y)$, contrast $\chi$, and entropy $\epsilon$ information (as suggested in [5]). We obtained one feature signature for every single image, where the signatures varied in size between 12 and 48 feature representatives. On average, a feature signature consisted of 30 representatives (i.e., 240 numbers per signature).

We combine individual descriptors by a (weighted) sum of their respective distances. As the individual descriptors (including the feature signatures) with the described distance functions form metric spaces, the combined space is also metric, which is important for efficient indexing. Individual distance components can be normalized and weighted as we will see further in Section 3.2.

**Querying and Evaluation Metrics.** Effectiveness of individual descriptors and their combinations is evaluated by precision of query answers. An image in the answer is considered part of precise answer, if it belongs to the same image class as the query image. Namely, we executed: 1,000 queries for ALOI dataset (each query from one image class), 100 queries for Corel Wang dataset (ten from each of the ten topics) and 200 queries for TWIC dataset (each from one image class). Within each dataset, we calculated *mean average precision* (MAP) [22] over the set of queries and also *average precision* for $k$-NN results with variable $k$. Further, we measured *intrinsic dimensionality* (iDIM) from the distance distribution of respective descriptor space [8].

## 3.2 Effectiveness of Individual Approaches

In this section, we describe the effectiveness of individual descriptors on all three datasets under test. Table 1 summarizes the results for all five global MPEG-7 descriptors individually and for the feature signatures with the SQFD distance.

As expected, the overall values of MAP differ significantly for individual datasets: ALOI is relatively uncomplicated dataset with MAP reaching values over 0.7 even

---

[1] The result contained many visually similar images.

**Table 1.** Mean Average Precision (MAP) of individual descriptors and their intrinsic dimensionality (iDIM)

| Individual descriptors | ALOI | | Corel Wang | | TWIC | |
|---|---|---|---|---|---|---|
| | MAP | iDIM | MAP | iDIM | MAP | iDIM |
| Color Layout | 0.37 | 1.3 | 0.43 | 4.2 | 0.17 | 4 |
| Color Structure | 0.78 | 2 | 0.5 | 7.7 | 0.20 | 6 |
| Edge Histogram | 0.28 | 2 | 0.4 | 6.8 | 0.15 | 5 |
| Region Shape | 0.15 | 0.7 | 0.25 | 2.4 | 0.07 | 2.5 |
| Scalable Color | 0.70 | 2.7 | 0.48 | 9.2 | 0.15 | 7 |
| SQFD | 0.73 | 3.7 | 0.49 | 15 | 0.32 | 18.5 |
| MPEG7 combination | 0.78 | 3 | 0.57 | 13 | 0.30 | 13.7 |

for single descriptors; we can observe MAP up to 0.5 for Corel Wang; and TWIC (as the most realistic dataset) has MAP values up to 0.2 for single global descriptors and 0.32 for feature signatures with SQFD. The last row of the table shows results for five combined MPEG-7 descriptors, each normalized by its maximum distance and summarized. As expected, this measure outperforms individual descriptors and, for ALOI and Corel Wang, it is better than SQFD.

The second columns for each dataset depict intrinsic dimensionality for the respective descriptor spaces. We can see that ALOI descriptors have significantly smaller iDIM which is caused by small actual visual difference between images in the dataset (all images depict single object isolated on a black background). Also, individual iDIM values often correspond with the effectiveness (MAP) of respective descriptors. As the feature signatures with SQFD cover several low-level features, the iDIM of this space is by far the highest.

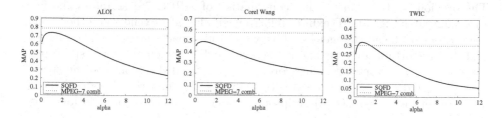

**Fig. 4.** MAP using global descriptor combination and SQFD with variable $\alpha$

As discussed in Section 2.2, SQFD allows to control the precision-indexability tradeoff by the parameter $\alpha$ used in the similarity function $f_s(r_i, r_j) = e^{-\alpha L_2(r_i,r_j)^2}$. In Figure 4, we can observe the MAP results for varying $\alpha$ in all three datasets. The optimal value of all of them is around value 1, which is caused by the fact that the feature extraction method was used with the same parameters. Too small or too big $\alpha$ results in less diverse values in the similarity matrix[2] and

---

[2] In the limit case, the resulting similarity matrix can be either nearly diagonal or unitary.

thus in the loss of information useful for similarity. Setting $\alpha = 1$ was used for all other experiments in this section (also in Table 1). For comparison, the figure also shows MAP of the combination of global MPEG-7 descriptors mentioned above.

We can summarize these results as follows: Effectiveness of the feature signatures with SQFD is mostly better than a single MPEG-7 global descriptor and comparable with the MPEG-7 descriptor combination. The question remains, whether better results can be reached by fusion of both approaches.

**Table 2.** Mean Average Precision (MAP) of various descriptor combinations: the distance measure of the combination is either (1) pure sum of its component sub-distances, or (2) sum of these sub-distances weighted by the components iDIM

| Effectiveness (MAP) | | | | | | |
|---|---|---|---|---|---|---|
| | ALOI | | Corel Wang | | TWIC | |
| Combination of descriptors | sum | iDIM | sum | iDIM | sum | iDIM |
| MPEG7 combination | 0.78 | 0.80 | 0.57 | 0.58 | 0.30 | 0.30 |
| Color Layout + SQFD | 0.71 | 0.74 | 0.50 | 0.51 | 0.33 | 0.34 |
| Color Structure + SQFD | 0.82 | 0.80 | 0.56 | 0.55 | 0.34 | 0.34 |
| Edge Histogram + SQFD | 0.71 | 0.74 | 0.52 | 0.52 | 0.35 | 0.34 |
| Region Shape + SQFD | 0.70 | 0.73 | 0.45 | 0.49 | 0.29 | 0.32 |
| Scalable Color + SQFD | 0.83 | 0.83 | 0.55 | 0.54 | 0.33 | 0.34 |
| MPEG7 combination + SQFD | 0.81 | **0.83** | 0.58 | **0.59** | 0.37 | **0.38** |

## 3.3   Approach Combinations

The previous section showed that combination of MPEG-7 descriptors and feature signatures with SQFD exhibit relatively similar effectiveness and their combination might be advantageous. All the descriptor spaces in question are metric and it is important for indexing and searching that the combination space preserve the metric properties. Therefore, we again decided to combine the spaces by a (weighted) sum of their respective distances. The first row of Table 2 shows again results of the MPEG-7 descriptor combination and then combinations with SQFD are presented.

For each dataset, the first column always means the pure sum of respective descriptors (normalized by maximum distances in each descriptor space), i.e.

$$\delta_{D_1 + \cdots + D_n}(X, Y) = \sum_{i=1}^{n} \frac{\delta_{D_i}(X_{D_i}, Y_{D_i})}{\text{max\_dist}_{D_i}}, \tag{1}$$

where $D_1, \ldots, D_n$ denote individual descriptors (e.g. Color Layout or feature signatures with SQFD), $X_{D_i}$ and $Y_{D_i}$ denote values of the $D_i$ descriptor of images $X$ and $Y$, respectively, and $\delta_{D_i}$ is the distance function used with descriptor $D_i$ (see Sections 2.1 and 2.2). When we compare MAP values from Tables 1 and 2, practically any combination (Table 2) reached higher MAP than its components

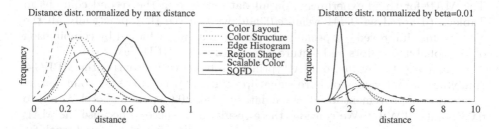

**Fig. 5.** Distance distribution of individual descriptors from TWIC: (a) normalized by maximum distance, (b) normalized by $\beta = 0.01$

(Table 1); the exceptional combinations (that worsened the MAP) are emphasized by italic numbers in Table 2. These exceptions always involve descriptors with extremely poor MAP (from Table 1).

A recent work [1] addressed the question of weight selection for combinations of metric visual descriptors. The authors studied the distance distribution of descriptor components normalized by the maximum distances to interval $[0, 1]$ – see distance histogram in Figure 5 (left). In general, descriptors with larger average distances would influence the combination sum more significantly, which might be a potential issue. We can try to overcome this by normalizing each descriptor $D_i$ by a distance $\tau_{D_i}$ smaller than maximum $\text{max\_dist}_{D_i}$. Looking at the distance histograms, the authors proposed to determine $\tau$ for each descriptor so that it corresponds to a certain fixed *percentage* $\beta$ of the smallest distances in the histogram (see [1] for details). The effect of this normalization for $\beta = 1\%$ is depicted in Figure 5 (right) – the beginnings of individual curves are very close, which should improve the effectiveness of the combination 5. Following this idea [1], we repeated all experiments with this normalization using $\beta = 1\%$ and $\beta = 1‰$, but the results were always slightly *worse* than with normalization by maximum distance. We plan to investigate this area even deeper in the future.

Nevertheless, we successfully applied another weight-tuning technique to improve the overall effectiveness of descriptor combination. As we mentioned already, we can observe a correlation between effectiveness of individual descriptors and their intrinsic dimensionality iDIM (see Table 1). In general, the iDIM tries to quantify the complexity of the data space and the difficulty to index such dataset using a metric access method [8]. The observed iDIM-MAP correlation can be naturally explained so that more complex descriptors have higher iDIM and their search effectiveness is higher. Barrios et al. made similar observation and they determine individual descriptor weights so that the iDIM of the combined space is maximized (finding a local maxima) [1]. We propose an alternative approach that builds directly on the observed correlation – we weight individual descriptors in the combination (1) by the respective iDIM:

$$\delta_{D_1 + \cdots + D_n}^{\text{iDIM}}(X, Y) = \sum_{i=1}^{n} \text{iDIM}_{D_i} \cdot \frac{\delta_{D_i}(X_{D_i}, Y_{D_i})}{\text{max\_dist}_{D_i}}. \tag{2}$$

The MAP for these experiments on all datasets are in the second columns in Table 2 (denoted as *iDIM weighted*) and we can see that practically all the MAP values improved – especially the overall maxima achieved by combination of all global descriptors and feature signatures with SQFD (last row).

Comparing these values of MAP (printed in bold) with values of the MPEG-7 combination and values of pure signatures with SQFD, this best combination resulted in improvement for all three datasets from 0.02 points (Corel Wang) to 0.08 points (TWIC). We consider these results as a success, because the MAP measure is always very difficult to improve, especially when the ground truth for each query forms only a small fraction of the whole collection (it is 0.5 % for the TWIC dataset, on average).

**Table 3.** Mean Average Precision (MAP) of searching $k$-NN using MPEG-7 desc. combination and then re-ranking of the kNN results using combination of MPEG-7 descriptors and signatures with SQFD

| Effectiveness (MAP) of SQFD re-ranking | | | | | | |
|---|---|---|---|---|---|---|
| | ALOI | | Corel Wang | | TWIC | |
| Combination of descriptors | sum | iDIM | sum | iDIM | sum | iDIM |
| MPEG7 100NN, re-rank MPEG7+SQFD | 0.79 | 0.81 | 0.44 | 0.45 | 0.28 | 0.28 |
| MPEG7 200NN, re-rank MPEG7+SQFD | 0.80 | 0.82 | 0.52 | 0.53 | 0.32 | 0.32 |
| MPEG7 300NN, re-rank MPEG7+SQFD | 0.81 | 0.82 | 0.55 | 0.56 | 0.33 | 0.33 |
| MPEG7 500NN, re-rank MPEG7+SQFD | 0.81 | 0.83 | 0.57 | 0.58 | 0.35 | 0.35 |
| MPEG7 1000NN, re-rank MPEG7+SQFD | 0.81 | 0.83 | – | – | 0.36 | 0.37 |

### 3.4   Re-Ranking by SQFD

As mentioned in Section 2.3, feature signatures with SQFD form a significantly more difficult data space for indexing and searching than MPEG-7 global descriptors accompanied with relatively cheap $L_p$-based distances. For larger datasets, it could be very time consuming to index the collection according to our most effective combination – MPEG-7 descriptors and feature signatures with SQFD.

This led us to the following schema: We index the data by the MPEG-7 descriptor combination, evaluate the $k$-NN(q) on such index, and re-rank these $k$ images according to distance "MPEG-7 combination + SQFD" evaluated with respect to query image $q$. Results of this approach are summarized in Table 3 for $k = 100, 200, 300, 500$, and $1,000$ – again for variants with individual weights equal to 1 and to respective iDIM values.

When we compare these results with the last row of Table 2 (MPEG-7 combination + SQFD), we can see that with growing $k$, the MAP gets very close to these maximal values. We can conclude, that even re-ranking by combination of MPEG-7 descriptors and feature signatures can improve the overall quality of the visual search almost as doing the whole search directly by the combination. However, the direct approach increases the search costs insignificantly.

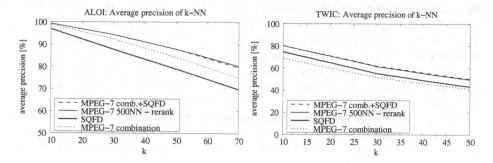

**Fig. 6.** ALOI and TWIC: Average precision of $k$-NN results for different approaches

The MAP is a very complex measure, but it does not necessarily give a good intuition of how good would be results of a real $k$-NN for a reasonable $k$; especially for larger datasets, improving MAP can be difficult even though precision of standard $k$-NN answers can be very high. Therefore, we measured also the $k$-NN *average precision* for variable $k$ and for the most important search approaches introduced above. See Figure 6 for these results on ALOI and TWIC datasets. Always, we present results for the combination of MPEG-7 descriptors, feature signatures with SQFD, the MPEG-7 combination + SQFD, and MPEG-7 500-NN re-ranked by MPEG-7 combination + SQFD. We can see that for ALOI, the precision is practically 100 % for 10-NN and falls down only to 80 % for 70-NN using the best approaches. Improvement of our combination in comparison with pure SQFD is up to 10 %. We can also see, that the average precision of the re-ranking approach is practically identical as the MPEG-7+SQFD approach. Note that the $x$-axis ends at 70 because that is the size of image classes for ALOI (analogously for other figures). For Corel Wang and TWIC dataset, the average precision is between 80 % and 50 % and the improvement of the combinations with respect to individual approaches is also about 10 %.

## 4   Conclusions and Future Work

In this paper, we combined feature signatures with MPEG-7 global visual descriptors to improve the performance of content-based image search engines. We proposed several techniques on how to combine these two orthogonal approaches and experimentally shown the positive synergistic effect of the combination. Hence, we could conclude that it is profitable to utilize both, MPEG-7 descriptors and feature signatures, because they complement each other and improve the quality of the retrieval. We also introduced a new (and more realistic) test image collection providing ground truth comprising images obtained via Google image search. The collection was designed to fill the gap in the realistic multimedia test collections with ground truth. Finally, we proposed a re-ranking variant guaranteeing efficient yet effective image retrieval. In the future, we plan to investigate feature signature extraction that profits from the descriptor combination. Also, we plan to compare our approach with the Bag-of-Words method as another standard approach successfully applied in multimedia retrieval. Additional experiments on larger datasets might show that MPEG-7 performance deteriorate faster than the performance of feature signatures. Hence, our re-ranking approach could contribute to more robust behavior on large databases.

Another interesting theoretical problem to inspect is the SQFD behavior under varying alpha parameter. It seems that there is an optimal value of alpha parameter for the precision, while with lower alpha we get always lower intrinsic dimensionality (iDIM is monotonously dependent on alpha). The relation between alpha parameter and the intrinsic dimensionality should be theoretically explained to provide more clues for the design of the SQFD distance spaces.

**Acknowledgments.** This research has been supported in part by Czech Science Foundation (GACR) projects P202/11/0968, P202/12/P297, 103/10/0886 and P202/10/P220.

# References

1. Barrios, J.M., Bustos, B.: Automatic weight selection for multi-metric distances. In: Proceedings of the Fourth International Conference on SImilarity Search and APplications, SISAP 2011, pp. 61–68. ACM Press, New York (2011)
2. Batko, M., Falchi, F., Lucchese, C., Novak, D., Perego, R., Rabitti, F., Sedmidubsky, J., Zezula, P.: Building a web-scale image similarity search system. Multimedia Tools and Applications 47(3), 599–629 (2010)
3. Batko, M., Kohoutkova, P., Novak, D.: CoPhIR Image Collection under the Microscope. In: Second International Workshop on Similarity Search and Applications (SISAP 2009), pp. 47–54. IEEE (2009)
4. Beecks, C., Lokoč, J., Seidl, T., Skopal, T.: Indexing the signature quadratic form distance for efficient content-based multimedia retrieval. In: Proc. ACM Int. Conf. on Multimedia Retrieval, pp. 24:1–24:8 (2011)
5. Beecks, C., Uysal, M., Seidl, T.: A comparative study of similarity measures for content-based multimedia retrieval. In: 2010 IEEE International Conference on Multimedia and Expo (ICME), pp. 1552–1557. IEEE (2010)
6. Beecks, C., Uysal, M.S., Seidl, T.: Signature quadratic form distance. In: Proc. ACM CIVR, pp. 438–445 (2010)
7. Beecks, C., Uysal, M.S., Seidl, T.: Signature quadratic form distance. In: Proc. ACM International Conference on Image and Video Retrieval, pp. 438–445 (2010)
8. Chávez, E., Navarro, G., Baeza-Yates, R., Marroquín, J.L.: Searching in metric spaces. ACM Computing Surveys 33(3), 273–321 (2001)
9. Coimbra, M.T., Cunha, J.P.S.: MPEG-7 Visual Descriptors -Contributions for Automated Feature Extraction in Capsule Endoscopy. IEEE Transactions on Circuits and Systems for Video Technology 16(5), 628–637 (2006)
10. Datta, R., Joshi, D., Li, J., Wang, J.Z.: Image retrieval: Ideas, influences, and trends of the new age. ACM Computing Surveys 40(2), 1–60 (2008)
11. Deb, S.: Multimedia Systems and Content-Based Image Retrieval. Information Science Publ. (2004)

12. Deselaers, T., Keysers, D., Ney, H.: Features for image retrieval: an experimental comparison. Information Retrieval 11(2), 77–107 (2008)
13. Eidenberger, H.: How good are the visual MPEG-7 features. In: SPIE Visual Communications and Image Processing Conference, vol. 5150, pp. 476–788. SPIE (2003)
14. Esuli, A.: PP-Index: Using permutation prefixes for efficient and scalable approximate similarity search. In: Proceedings of LSDS-IR 2009 (2009)
15. Geusebroek, J.-M., Burghouts, G.J., Smeulders, A.W.M.: The Amsterdam Library of Object Images. IJCV 61(1), 103–112 (2005)
16. Hafner, J., Sawhney, H.S., Equitz, W., Flickner, M., Niblack, W.: Efficient color histogram indexing for quadratic form distance functions. IEEE Transactions on Pattern Analysis and Machine Intelligence 17, 729–736 (1995)
17. Hu, R., Rüger, S., Song, D., Liu, H., Huang, Z.: Dissimilarity measures for content-based image retrieval. In: Proc. IEEE International Conference on Multimedia & Expo., pp. 1365–1368 (2008)
18. Kruliš, M., Lokoč, J., Beecks, C., Skopal, T., Seidl, T.: Processing the signature quadratic form distance on many-core gpu architectures. In: Proceedings International Conference on Information and Knowledge Management, pp. 2373–2376 (2011)
19. Lokoč, J., Hetland, M., Skopal, T., Beecks, C.: Ptolemaic indexing of the signature quadratic form distance. In: Proceedings of the Fourth International Conference on SImilarity Search and Applications, pp. 9–16. ACM (2011)
20. Lokoč, J., Novák, D., Skopal, T., Sibirkina, N.: Thematic Web Images Collection. SIRET Research Group (2012), http://siret.ms.mff.cuni.cz/twic
21. Manjunath, B.S., Salembier, P., Sikora, T. (eds.): Introduction to MPEG-7: Multimedia Content Description Interface. John Wiley & Sons, Inc., New York (2002)
22. Manning, C.D., Raghavan, P., Schütze, H.: Introduction to information retrieval. Cambridge University Press (2008)
23. Mikolajczyk, K., Schmid, C.: A performance evaluation of local descriptors. IEEE Transactions on Pattern Analysis and Machine Intelligence 27(10), 1615–1630 (2005)
24. MPEG-7. Multimedia content description interfaces. Part 3: Visual. ISO/IEC 15938-3:2002 (2002)
25. Novak, D., Batko, M., Zezula, P.: Metric Index: An efficient and scalable solution for precise and approximate similarity search. Information Systems 36(4), 721–733 (2011)
26. Puzicha, J., Buhmann, J.M., Rubner, Y., Tomasi, C.: Empirical evaluation of dissimilarity measures for color and texture. In: Proc. IEEE International Conference on Computer Vision, vol. 2, pp. 1165–1172 (1999)
27. Skopal, T., Bartoš, T., Lokoč, J.: On (not) indexing quadratic form distance by metric access methods. In: Proc. Extending Database Technology (EDBT). ACM (2011)
28. Spyrou, E., Le Borgne, H., Mailis, T., Cooke, E., Avrithis, Y., O'Connor, N.: Fusing MPEG-7 visual descriptors for image classification, pp. 847–852 (September 2005)
29. Stanchev, P., Amato, G., Falchi, F., Gennaro, C., Rabitti, F., Savino, P.: Selection of MPEG-7 image features for improving image similarity search on specific data sets. In: 7th IASTED International Conference on Computer Graphics and Imaging, CGIM, pp. 395–400. Citeseer (2004)
30. Wang, J.Z., Li, J., Wiederhold, G.: Simplicity: Semantics-sensitive integrated matching for picture libraries. IEEE Transactions on Pattern Analysis and Machine Intelligence 23, 947–963 (2001)

# Revisiting Techniques for Lowerbounding the Dynamic Time Warping Distance

Tomáš Bartoš and Tomáš Skopal

SIRET Research Group,
Faculty of Mathematics and Physics,
Department of Software Engineering,
Charles University in Prague, Czech Republic
{bartos,skopal}@ksi.mff.cuni.cz
http://siret.cz

**Abstract.** The dynamic time warping (DTW) distance has been used as a popular measure to compare similarities of numeric time series because it provides robust matching that recognizes warps in time, different sampling rate, etc. Although DTW computation can be optimized by dynamic programming, it is still expensive, so there have been many attempts proposed to speedup DTW-based similarity search by distance lowerbounding. Some approaches assume a constrained variant of DTW (i.e., fixed dimensions, warping window constraint, ground distance), while others do not. In this paper, we comprehensively revisit the problem of DTW lowerbounding, define a general form of DTW that fits all the existing variants and goes even beyond. For the constrained variants of general DTW we propose a lowerbound construction generalizing the LB_Keogh that for particular ground distances offers speedup by up to two orders of magnitude. Furthermore, we apply metric and ptolemaic lowerbounding on unconstrained variants of general DTW that beats the few existing competitors up to two orders of magnitude.

## 1 Introduction

The similarity search in databases of time series is a popular technique used in data mining, multimedia retrieval, and other domains. In fact, the time series is a suitable content-based descriptor for data types where the extracted features, i.e., the elements of time series, need to be treated in a certain order (not necessarily interpreted as the time). Therefore, a robust similarity between time series should take into account the order of time series elements, interpreting the neighboring elements more or less similarly. Hence, vectorial distances, such as $L_p$ distances, are not appropriate in this case as they treat the time series elements (dimensions) independently. There have been many distances proposed in the time series retrieval domain, while they all provide some *alignment* between two time series based on an optimization criterion. One of the most popular distances is the dynamic time warping distance (DTW).

G. Navarro and V. Pestov (Eds.): SISAP 2012, LNCS 7404, pp. 192–208, 2012.

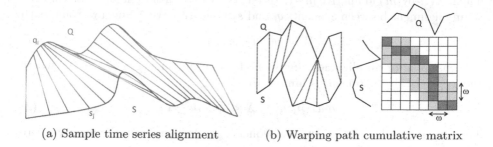

(a) Sample time series alignment    (b) Warping path cumulative matrix

**Fig. 1.** Dynamic Time Warping distance

Despite the success of DTW demonstrated by applications in many domains, we find a lot of inconsistencies over the available literature where efficient similarity search under DTW by use of lowerbounding is discussed. The definition of DTW problem itself varies substantially, so that individual contributions to DTW-based similarity search are not always comparable. In particular, there are different assumptions on the time series lengths (fixed or variable), the ground distance used (specific or general), and the presence/type of warping constraints.

In this paper, we unify the mosaic of existing approaches to DTW lowerbounding into a consistent framework that allows to recognize them in a fair way. For this purpose, we define a general form of DTW that fits all the variants used in the literature (constrained and unconstrained) and goes even beyond. Moreover, for the constrained variants of general DTW we propose a new lower bound construction generalizing LB_Keogh method that for particular ground distances offers speedup by an order of magnitude. Furthermore, we apply pivot-based metric and ptolemaic lowerbounding on unconstrained variants of general DTW, showing that this approach beats the existing competitors several times.

## 2    Dynamic Time Warping Distance

One of the very first definitions of *Dynamic Time Warping distance* [2] declares DTW as a technique for aligning time series with a predefined word template. It has been proposed primarily for speech recognition processes with the main goal to minimize the dissimilarity between a speech sample and speech patterns to find a "perfect" match. Over time, DTW has become a popular technique for measuring the similarity between general time series. In particular, DTW has an ability to "elastically" align two time series, which leads to robust matching that recognizes warps in time, different sampling rate, etc.

Speaking more precisely, the initial definition of DTW distance function is as follows: we take two time series $Q$ (query object) and $S$ (database object) of lengths $n$ and $m$ each of which contains a number of inner elements

$$Q = q_1, q_2, \ldots, q_i, \ldots, q_n \quad \text{and} \quad S = s_1, s_2, \ldots, s_j, \ldots, s_m$$

Then, an $n$-by-$m$ cumulative matrix is created where each item $(i, j)$ corresponds to an alignment between elements $q_i$ and $s_j$ (Fig. 1a). We define a warping path

$$W = w_1, w_2, \ldots, w_X, \tag{1}$$

as a mapping or an alignment between elements in time series $Q$ and $S$ where the length of the alignment is limited to

$$max(n, m) \leq X < n + m - 1 \tag{2}$$

The warping path (see Fig. 1b) is a sequence of elements from the matrix where each element $w_k = (i, j)_k$ must meet several criteria, namely

- **Boundary conditions** to restrict the searching space for warping paths
- **Monotonicity** The element pairing is monotonous according to the time
- **Continuity** to ensure that neighboring elements in the warping path correspond to adjacent cells within the matrix.

Finally, the DTW is defined as a minimized warping path from the universe of all acceptable warping paths as

$$DTW(Q, S, \delta) = min_W \left\{ \sum_{k=1}^{X} \delta(w_k) \right\} \tag{3}$$

where $\delta$ is a ground distance (the distance measure between two elements of time series). Generally, this might be any distance but we suppose numeric time series, so the commonly considered distance functions include $\delta_1(i, j) = |q_i - s_j|$ or $\delta_2(i, j) = |q_i - s_j|^2$. These $\delta_p(\cdot, \cdot)$ functions might be viewed as simplified $L_p$ distance measures where $L_p$ stands for family of Minkowski distances [19].

Because there is an exponential number of possible warping paths, we employ a method of dynamic programming which is often used for evaluating DTW [14].

## 2.1   Constrained DTW

Some domains such as speech recognition require alignments between time series to have some additional constraints – to avoid warping paths with excessive time stretch, to avoid sequence distortions, or to discard "non-interesting" warping paths [13]. These include

- **Slope constraint** is an example of a local constraint which restricts the slope of warping paths in order to limit very large movements of the warping path in a single direction. In [14] authors suggest four types of such a constraint.
- **Warping window** To define a warping window $\omega$ in which elements of the warping path must fit. For a positive integer $\omega$, values $w_{k-1} = (a, b)$ and $w_k = (\overline{a}, \overline{b})$, we get the condition $\overline{a} - a \leq \omega$. For our purposes, we focus on Sakoe-Chiba band [14] and we will study the DTW indexability depending on the different values of $\omega$. In Fig. 1b see the warping window $\omega = 2$ as the light gray area.

**Table 1.** Basic notation

| Symbol | Description |
|---|---|
| $Q, S$ | query and database time series (e.g., numeric time series) |
| $\delta_p$ or $\delta_p(\cdot,\cdot)$ | simplified $L_p$ distance between two elements of time series |
| DTW $(\cdot,\cdot,\delta)$ | dynamic time warping distance with ground distance function $\delta$ |
| GDTW $(\cdot,\cdot,\delta,\omega,f)$ | generalized DTW$(\cdot,\cdot,\delta)$ with warping window constraint $\omega$ and a monotonic function $f$ |
| LB_{name} | lowerbounding technique |
| LB_Pivots (Met) | pivot-based lowerbounding (metric) |
| LB_Pivots (Pto) | pivot-based lowerbounding (ptolemaic) |

**Table 2.** Comparison of lowerbounding techniques

| Method | DTW fixed length | | | | DTW variable length | | | |
|---|---|---|---|---|---|---|---|---|
| | $\delta_1$ | $\delta_2$ | $\delta_p$ | $\delta_{LB}$ | $\delta_1$ | $\delta_2$ | $\delta_p$ | $\delta$ |
| LB_Keogh | ✓ | ✓ | ✓ | n/a | | | | |
| LB_Kim | ✓ | | | | ✓ | | | |
| LB_Yi | ✓ | | | | ✓ | | | |
| LB_Tight | ✓ | ✓ | ✓ | ✓ | | | | |
| LB_Pivots(Met) | ✓ | ✓ | ✓ | ✓ | ✓ | ✓ | ✓ | ✓ |
| LB_Pivots(Pto) | ✓ | ✓ | ✓ | ✓ | ✓ | ✓ | ✓ | ✓ |

## 2.2 Generalized DTW

In the past, there appeared several publications with slightly different definitions of DTW (e.g., [2,7,18]) in which authors stated their definition to be the (only) one valid to fit their needs. To clarify and generalize all the mentioned definitions, we define the generalized DTW distance for an arbitrary ground distance

$$GDTW(Q, S, \delta, \omega, f) = min_W \left\{ f(\sum_{k=1}^{X} \delta(w_{k,\omega})) \right\} \tag{4}$$

where $w_{k,\omega}$ is $k$th item in the working path $W$ with the corresponding warping window constraint $\omega$ and $f$ is a monotonic function. Some authors [7] use $f = \sqrt{\cdot}$ because if we take the zero warping window together with ground distance $\delta_2$, the result distance corresponds to $L_2$ (Euclidean) distance

$$GDTW(\cdot, \cdot, \delta_2, 0, \sqrt{\cdot}) = L_2(\cdot, \cdot) \tag{5}$$

For completeness, Table 1 displays the notation that we use in the following text.

## 3 Related Work

Even if we use the dynamic programming method [14], the time complexity of a single DTW distance computation between arbitrary time series with lengths $n, m$ remains in $O(n \cdot m)$. So, there have been many efforts spent on speeding the similarity search under DTW, varying in the assumptions on the DTW variant used (e.g., constrained or unconstrained). Generally, all the approaches utilize the concept of *lowerbounding* – a "cheaper" distance function that eliminates non-interesting objects to avoid as many "expensive" DTW computations as possible.

The lowerbounding concept works with an estimation of DTW distance which always gives a value smaller or equal to a result of the DTW computation, without the need of computing the DTW itself:

$$DTW(Q, S, \delta) \geq \texttt{LowerBound}(Q, S) \qquad (6)$$

If the lowerbounding considers $\delta$, it usually gives tighter lowerbound values which results in better object filtering for the most popular similarity queries such as range or $k$NN queries. Table 2 displays the comparison between three selected and existing lowerbounding schemes in the upper part, together with our three proposed techniques that will be described in Section 4 in the lower part. The table shows which techniques are applicable for time series of fixed or variable lengths, and which support a particular ground distance $\delta$.

### 3.1   Early Abandoning

The simplest idea of speeding up the DTW distance computations in the context of similarity search is to use the *early abandoning* method. Given a radius $r_Q$ of a query that is being evaluated[1], we can stop the DTW computation if we know that the final distance will be greater than $r_Q$ [6]. This eliminates further computations of the DTW matrix, discards inappropriate objects early enough, and provides not aproximate but exact results. In fact, the early abandoning makes DTW a lowerbounding function to itself.

One suggested technique is to focus on dynamic programming, cumulative distances $\gamma(i, j)$, and their values [6]. We can eliminate further calculations when either (a) all adjacent cells exceed $r_Q$ or (b) all cells in a row/column exceed $r_Q$.

As this method gives faster yet still correct results and because its usage is orthogonal to the other approaches, we further consider early abandoning in all other lowerbounding methods (if not stated otherwise). For completeness, there exists another orthogonal method (the "anticipatory" pruning) that might be further applicable [1] to improve efficiency of DTW computations.

### 3.2   Time Series-Specific Lowerbounding

In the following text, we will study several existing methods that introduce different concepts of how to obtain specific lowerbounds and name them according to their authors as proposed in [7]. As we mentioned before, authors usually do not take into account the same definitions of DTW, so we compare these approaches between each other focusing on the GDTW definition (see Eq. 4).

**LB_Yi.** In one of the first papers focusing on DTW lowerbounding the authors suggested a technique known as LB_Yi [18] that computes a lowerbound distance

---

[1] Either the fixed radius of a range query, or the current radius of a $k$NN query.

for two time series $Q, S$ depending on the arrangements of their ranges defined as $[min(Q), max(Q)]$ and $[min(S), max(S)]$. If the lengths of time series are $m$ and $n$, respectively, we get the lowerbound in $O(m+n)$. Moreover, the advantage of this technique is its ability to treat time series of variable lengths.

Authors claim that this method works for DTW with ground distances $\delta_1$ and $\delta_2$ but, as we point out, this is not true for $\delta_2$ because of an incorrect observation $|max(Q) - max(S)| \leq DTW(Q, S, \delta_2)$. The reason for this might be the inconsistency in the definition of $\delta_2$ – although authors define it generally as $\delta_p(x, y) = \sum_{i=1}^{n} |x_i - y_i|^p$ (which for $p = 2$ is squared $L_2$ distance), they call it Euclidean distance. This results in the inapplicability of LB_Yi as a method for lowerbounding $DTW(\cdot, \cdot, \delta_2)$. For completeness, we also provide the counter example of incorrect lowerbound.

*Example 1.* Suppose we have $DTW(\cdot, \cdot, \delta_2)$ and time series with two real numbers $Q = \{-1,2\}$ and $S = \{-1.1, 1.9\}$. The ranges overlap, so the lower bound is

$$\texttt{LB\_Yi}(Q, S) = \sum_{q_i > max(S)} |q_i - max(S)| + \sum_{s_j < min(Q)} |s_j - min(Q)| \quad = 0.2$$

When we take these time series and compute DTW distance, we will see that the best alignment is on the diagonal with the result of

$$DTW(Q, S, \delta_2) = |-1 - (-1.1)|^2 + |2 - 1.9|^2 = 0.1^2 + 0.1^2 = 0.02$$

It is easy to see that the computed value of LB_Yi is greater than the actual value of DTW distance, so in this case LB_Yi cannot be used as a valid lowerbound.

**LB_Kim.** Another approach of estimating DTW is presented in the work [10]. According to authors, LB_Kim guarantees the lower bounds when the ground distance is any $L_p$ distance function which, in our terms, applies to any $DTW(\cdot, \cdot, \delta_p)$.

The mechanism is based on extracting and using 4-tuple vectors consisting of (1) first, (2) last, (3) smallest (min), and (4) largest (max) elements of two time series $Q, S$. To compute the lower bound value, authors consider the maximum value of all absolute values of differences

$$\texttt{LB\_Kim}(Q, S) = max(|q_1 - s_1|, |q_{last} - s_{last}|, |q_{min} - s_{min}|, |q_{max} - s_{max}|)$$

It is a simple observation, that if the time series are long, this typically does not give a very tight lowerbound. On the other hand, the method works with time series with different lengths. But, as we show in the following example, if the ground distance $\delta_2$ is used, the LB_Kim is not a guaranteed lowerbound.

*Example 2.* Suppose we have $DTW(\cdot, \cdot, \delta_2)$ with two time series $Q = \{2, 1\}$ and $S = \{1.9, 1.1\}$. It is easy to verify that the resulting DTW distance is 0.02. Having the previous lower bound definition, we get incorrect lower bound value

$$\texttt{LB\_Kim}(Q, S) = max(|2 - 1.9|, |1 - 1.1|, |1 - 1.1|, |2 - 1.9|) = 0.1$$

**LB_Keogh.** The best lowerbounding method for constrained DTW distance and fixed-length time series so far is considered LB_Keogh [7]. With warping window constraint $\omega$, this method encapsulates any input time series $S$ of length $n$ into two additionally computed time series $U_S$ and $L_S$ (upper and lower part):

$$u_i = max(s_{i-\omega}, \ldots, s_{i+\omega}), \quad l_i = min(s_{i-\omega}, \ldots, s_{i+\omega})$$

Given a query object $Q$ together with time series $U_S, L_S$ that correspond to Keogh's envelope for an input time series $S$ (all with the same length $n$), the proposed lower bound LB_Keogh is defined as

$$\text{LB\_Keogh}(Q, S) = \sqrt{\sum_{i=1}^{n} \begin{cases} (q_i - u_i)^2 & \text{if } q_i > u_i \\ (q_i - l_i)^2 & \text{if } q_i < l_i \\ 0 & \text{otherwise} \end{cases}}$$

Several experiments verified claims that this lowerbound is the tightest one compared to other approaches [7]. Unfortunately, this is true only for the particular definition of DTW distance given by the author (see Eq. 5). If we take the generalized DTW distance and replace the ground distance $\delta_2$ with other distance such as $\delta_1$, we could find a tighter lowerbound. We will describe this situation in more details in Section 4.

### 3.3    Pivot-Based Lowerbounding

If we consider DTW or GDTW (fixed or variable dimensionality, with or without warping window) as a black-box distance function $d$ that satisfies certain topological properties, we could utilize *pivot-based lowerbounding* instead of the time series-specific ones [12]. The pivot-based lowerbounding relies just on pre-computed DTW distances from query/database objects to so-called *pivots*.

– **Metric Lowerbounding.** When a distance function used for creating a pivot table obeys the metric postulates (i.e., *identity, positivity, symmetry,* and *triangle inequality*), we could leverage *metric* pivot-based lowerbounding, i.e., we could define a metric lowerbounding principle that is generally used within the whole class of so-called *metric access methods* [4,19].

– **Ptolemaic Lowerbounding.** Analogously, in *Ptolemaic indexing* [5,11], we construct lower bounds using *Ptolemy's inequality* [5,11].

**Is DTW metric or ptolemaic distance?** Because the pivot-based lowerbounding utilizes the pair-wise distances between individual objects, it could be used with any black-box distance that satisfies metric and/or ptolemaic properties. However, neither DTW nor GDTW does satisfy the triangle or Ptolemy's inequality, thus the metric and ptolemaic lowerbounding cannot be used directly. Nevertheless, in Section 4.2 we discuss a simple modification of GDTW that obeys the properties of metric and/or ptolemaic distance, enabling to use pivot-based lowerbounding for efficient similarity search under GDTW.

## 3.4   Advanced Indexing

There have also been proposed some advanced techniques for indexing DTW. In [18] authors suggest using the FastMap index, that maps the DTW distances into Euclidean space. This approach, however, leads to only approximate estimates of DTW using Euclidean distance, so exact searching cannot be guaranteed.

An enormous effort in speeding DTW-based similarity search has been done by Eamonn Keogh's group. In famous papers [7,8] introducing LB_Keogh, authors consider additional dimensionality reduction of the time series using piecewise aggregate approximation and indexing of the reduced envelopes by R-tree.

In this paper we revisit the "lower level" of DTW indexing – the lowerbounding techniques that determine the success of any advanced indexing techniques based on lowerbounding. Hence, we do not consider high-level indexing solutions such as iSAX framework [15,3]. Moreover, as LB_Keogh-based solutions assume only fixed lengths of time series, warping path constraints, and only $\delta_2$ ground distance, they cover only a fraction of scenarios we consider here.

# 4   Lowerbounding of Generalized DTW

Previous sections revealed some issues if the DTW definition is not unified. As we propose GDTW (see Eq. 4) as a generalized version of DTW, we also define a general lowerbounding technique applicable for arbitrary ground distances.

One of our objectives is to create a lowerbounding technique that adjusts to the ground distance used in GDTW in order to produce the best lowerbound compared to other methods that are defined generally and remain unchanged for different ground distances. If we restrict the type of ground distances in GDTW to some extent, we can propose a modified version of LB_Keogh that gives very good results in terms of lowerbound tightness. We label it as LB_Tight.

**Definition 1.** *Given a query object Q together with time series $U_S, L_S$ that correspond to Keogh's envelope for an input time series S (all with the same length n), using $\omega$ as the warping window parameter, the proposed lowerbound LB_Tight for GDTW$(Q, S, \delta, \omega, f)$ is defined as*

$$LB\_Tight(Q, S, \delta, f) = f \left( \sum_{i=1}^{n} \begin{cases} \delta(q_i, u_i) & \text{if } q_i > u_i \\ \delta(q_i, l_i) & \text{if } q_i < l_i \\ 0 & \text{otherwise} \end{cases} \right)$$

The definition is based on the LB_Keogh with very small modifications. If we take square root ($\sqrt{}$) as the monotonic function $f$ and the ground distance is $\delta_2$, our lowerbound definition equals to LB_Keogh. On the other hand, the proposed definition gives better applicability as it serves as a correct lowerbound for DTW with a wider range of ground distance functions than a single one such as $\delta_2$ as we will see in the following text.

## 4.1    LB_Tight as a General Lowerbounding Function for GDTW

To verify the correctness of our proposed lower bound, we need to limit the applicable ground distances that might be used to some extent - we denote such functions as *LB-compliant*. Afterwards, we show that for these functions LB_Tight is correctly defined lowerbound for GDTW.

**Definition 2.** *A distance function $\delta(x, y)$ is LB-compliant iff the inputs $x, y$ are real numbers and it fulfills the following requirements:*

*(a) $\delta(x, y)$ is **non-negative**: $\delta(x, y) \geq 0$*

*(b) $\delta(x, y)$ is **symmetric**: $\delta(x, y) = \delta(y, x)$*

*(c) for $x, y \in \mathbb{R}$, $\epsilon \in \mathbb{R}_0^+$ such that $x \geq y + \epsilon \geq y$ the following inequalities hold*

$$\delta(x, y + \epsilon) \leq \delta(x, y) \tag{7}$$
$$\delta(x + \epsilon, y) \geq \delta(x, y) \tag{8}$$

Now, we can easily observe that besides other functions, any distance function from the $\delta_p$ family fulfills the above criteria as stated in the Theorem 1. For functions that are *LB-compliant* and are used as ground distances in GDTW, our lowerbounding mechanism works as it is confirmed by Theorem 2.

**Theorem 1.** *Any $\delta_p$ function with $p \geq 1$ is LB-compliant.*

*Proof.* See Appendix A.1.

**Theorem 2.** *For any two time series $Q$ and $S$ of the same length $n$, for any value of the warping window $\omega$ such that $(j - \omega) \leq i \leq (j + \omega)$, and for any ground distance $\delta$ that is LB-compliant, the following inequality holds:*

$$\texttt{LB\_Tight}(Q, S, \delta, f) \leq GDTW(Q, S, \delta, \omega, f)$$

*Proof.* See Appendix A.2.

## 4.2    Pivot-Based Lowerbounding of Generalized DTW

As discussed in Section 3.3, a pivot-based lowerbounding cannot be directly used with DTW or GDTW, because they do not satisfy the metric or ptolemaic properties. In fact, they are *semi-metrics*, so they satisfy just *identity, positivity,* and *symmetry*. To turn GDTW into a metric (or ptolemaic) distance, we need to find a modification that generates the missing property yet preserves the original GDTW semantics (the similarity ordering of objects w.r.t. any query).

**Making GDTW a Metric.** The TriGen algorithm [16,17] was proposed to keep a user-controlled amount of triangle inequality in a semi-metric distance. Given a rate of non-triangular triplets (T-error tolerance), the TriGen algorithm finds a T-modifier for which the intrinsic dimensionality $\rho$ [4] is minimized, while the T-error does not exceed the tolerance. The modified distance $g(d)$ determined by TriGen can be then employed by the pivot table for exact but slower (T-error tolerance is zero, so $\rho$ gets higher) or approximate but fast (T-error tolerance is positive, so $\rho$ gets smaller) similarity search. For our purposes, we utilize TriGen to generate the full triangle inequality into GDTW (i.e., zero T-error tolerance) while using Fractional-Power (FP) base as the so-called *T-bases* [16,17].

**Making GDTW a Ptolemaic Metric.** Securing the Ptolemy's inequality seems more complicated, however, we can use a simple trick. As shown in [5], every metric $d$ can be made also ptolemaic by applying the square root, i.e., $\sqrt{d}$ is used instead of $d$. In the previous paragraph we utilized the TriGen algorithm to turn GDTW into a metric distance, hence, a further application of square root makes the GDTW a ptolemaic metric distance.

## 5   Experiments

In the previous text, we theoretically proved that not all existing lowerbounding techniques are suitable for GDTW distance because some of them strictly depend on the ground distance $\delta$. We also stated that LB_Keogh might not be the tightest lowerbound if the $\delta_1$ is used as the ground distance in DTW. To validate our statements, we evaluated the lowerbounding approaches on selected datasets.

### 5.1   Datasets

We wanted to run the experiments on real-world data, so we selected well-known *UCR Time Series* [9] collected by Keogh. In order to avoid cherry-picking but still limit the number of datasets, we tested only datasets with more than 100,000 numbers (the number of time series multiplied by the number of their elements). With this approach, we obtained 25 datasets of various types.

We added a dataset with time series of variable lengths that we obtained from a set of 20,660 DNA sequences of genes of *Listeria monocytogenes*[2]. using the technique suggested in [3]. Moreover, we created an additional dataset from the stock market *NASDAQ*[3] for which we took stock open and close prices between years 1970 and 2010 and we divided them into time series with lengths up to 1024. For better precision, we subtracted the mean value from each time series.

---

[2] For details see *Metric Spaces Library* at http://sisap.org/
[3] http://www.infochimps.com/datasets/nasdaq-exchange-daily-1970-2010-open
-close-high-low-and-volume

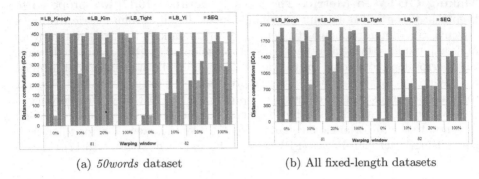

(a) *50words* dataset                    (b) All fixed-length datasets

**Fig. 2.** Average number of Distance Computations

## 5.2   Experimental Settings

For each dataset, we monitored how different lowerbounding methods behave for a specific ground distance $\delta$ in terms of (1) Distance computations (DCs), (2) Total query real time, and (3) Speedup in real time compared to the sequential scan ($\frac{Time_{SEQ}}{Time_{LB}}$). For clarification, we observed also the (4) Real error to confirm that some lowerbounding methods were not true lower bounds for all ground distances. We averaged our experiments for 10NN queries over 100 random query objects and tested ground distances $\delta_1$ and $\delta_2$ together with $\sqrt{}$ as the function $f$ (see Eq. 4).

For the fixed-length time series, we used several values of the Sakoe-Chiba band [14] which corresponds to warping window constraint $\omega$ . We tested 5 values of $\omega$ corresponding to 0%, 5%, 10%, 15%, and 20% of the query length (similar to [7]) together with the unconstrained DTW ($\omega = \infty$).

For time series of variable lengths (*Listeria* and *NASDAQ*), we tested only the unconstrained version of DTW. To apply both pivot-based lowerbounding methods LB_Pivots(Met) and LB_Pivots(Pto), we obtained the corresponding T-modifiers for a zero T-error tolerance using the TriGen algorithm with a sample database of 300 random objects from which we generated 100,000 triplets (including 1% of anomalous triplets).

## 5.3   Results

The following figures depict the results we obtained through all experiments. All lowerbounding methods use DTW with *early abandoning* (see Section 3.1) for better real time responses. An exception is the sequential scan, where the variant denoted as SEQ does not use early abandoning, while SEQ (EA) does. For transparency, we divided the outcomes in three sections according to the datasets – fixed-length time series, *Listeria*, and *NASDAQ*.

**Fixed-Length Datasets.** Tests confirmed that LB_Tight outperforms all other methods for ground distance $\delta_1$ in terms of the number of distance computations,

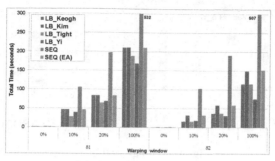

(a) Average query time

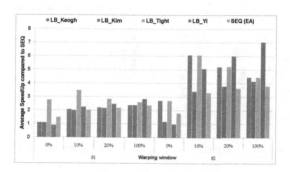

(b) Average Speed-up vs. SEQ scanning

**Fig. 3.** All fixed-length datasets

so it is the tightest lowerbound achieving up to two orders of magnitude speedup over competitors (Fig. 2a, Fig. 2b).

For ground distance $\delta_2$, LB_Keogh, which equals to LB_Tight$(\cdot, \cdot, \delta_2)$, remains the dominant lowerbounding mechanism. As the results were similar for most datasets, we explicitly show *50words* dataset (Fig. 2a) followed by average results over all fixed-length datasets (Fig. 2b).

Although the computations of lowerbounds generate some overhead, the combination with early abandon technique outperforms the sequential scan, yet enabling the exact search (Fig. 3a). We show the real speedup compared to pure sequential scanning in Fig. 3b. Here we see the speedup of up to 346% for LB_Tight with ground distance $\delta_1$ and up to 607% for LB_Keogh with ground distance $\delta_2$.

Interestingly, as the warping window enlarges, the tightness of LB_Tight lowerbounding decreases while LB_Yi lowerbounds improve. However, to be fair, we have to admit that at the same time, the error rate of LB_Yi increases (see Fig. 4a). LB_Keogh is not included in Fig. 4a as it worked correctly (with no false dismissals) for both ground distances.

**Listeria Dataset.** In case of *Listeria*, we compared our proposed methods LB_Pivots(Met), LB_Pivots(Pto) only with LB_Kim and LB_Yi, as other lowerbounding approaches were not valid for time series of variable lengths.

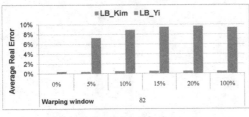

(a) All fixed-length datasets

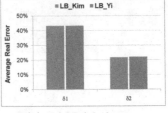

(b) NASDAQ dataset

**Fig. 4.** Average real error

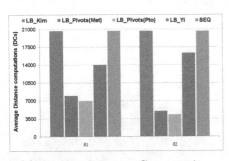

(a) Average Distance Computations

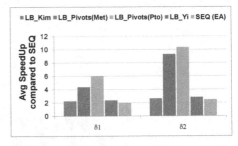

(b) Average Speed-up vs. SEQ scanning

**Fig. 5.** *Listeria* dataset

Here we show the dominance of our methods for lowerbounding DTW regardless of the ground distance function $\delta$. We begin with the number of DCs (Fig. 5a) where ptolemaic approach LB_Pivots(Pto) is slightly better than the metric LB_Pivots(Met). Yet both methods give a substantial decrease of DCs.

This results in the improvements of the total query time (Fig. 7a) also because of employed early abandon technique. Last but not least, there is a tremendous speedup of more than 6× for ground distance $\delta_1$ and up to 10.3× for ground distance $\delta_2$ (Fig. 5b). Together with the zero real error, we confirmed the applicability of suggested methods for lowerbounding the general DTW.

**NASDAQ Dataset.** *NASDAQ* dataset just confirms the previously achieved results. The suggested pivot-based methods LB_Pivots(Met) and LB_Pivots(Pto) give considerably good results for all monitored values such as DCs (Fig. 6a), total query time (Fig. 7b), speedup (up to 4.1× for ground distance $\delta_1$ and more than 8.7× for ground distance $\delta_2$; see Fig. 6b). Although LB_Yi was better in time efficiency, it turns out that the error rate was extremely high, up to 43% (Fig. 4b) which yields in inapplicability for exact searching.

### 5.4   Summary

All experiments confirmed proposed methods for lowerbounding as the best-of-breed techniques for exact similarity search under GDTW. The generalized

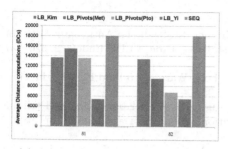

(a) Average Distance Computations

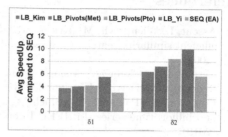

(b) Average Speed-up vs. SEQ scan

**Fig. 6.** *NASDAQ* dataset

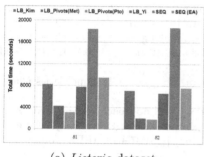

(a) *Listeria* dataset.

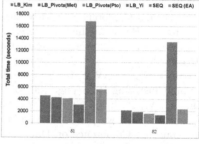

(b) *NASDAQ* dataset.

**Fig. 7.** Total query time

LB_Tight outperformed all other lowerbounding approaches in the area of fixed-length datasets. In *Listeria* and *NASDAQ* datasets both lowerbounding methods, metric LB_Pivots(Met) and ptolemaic LB_Pivots(Pto), improved the speed of SEQ scanning of up to 10.3× with no false dismissals.

## 6 Conclusions

In this paper we revisited the problem of DTW lowerbounding and defined a general form of DTW that unified all the variants used in literature (including even more general cases). For the constrained variants of general DTW we have proposed a novel lowerbound construction generalizing the LB_Keogh approach, while in the experimental results it demonstrated a speedup by up to two orders of magnitude (for particular ground distances). Furthermore, we proposed an alternative approach of lowerbounding unconstrained variants of general DTW based on pivot-based metric and ptolemaic lowerbounds, while these techniques outperformed the few existing competitors by up to two orders of magnitude.

**Acknowledgments.** This research has been supported by Czech Science Foundation (GAČR) project 202/11/0968 and by Grant Agency of Charles University (GAUK) project 567312.

# References

1. Assent, I., Wichterich, M., Krieger, R., Kremer, H., Seidl, T.: Anticipatory dtw for efficient similarity search in time series databases. Proc. VLDB Endow. 2, 826–837 (2009)
2. Berndt, D.J., Clifford, J.: Using Dynamic Time Warping to Find Patterns in Time Series. In: KDD Workshop, pp. 359–370 (1994)
3. Camerra, A., Palpanas, T., Shieh, J., Keogh, E.: iSAX 2.0:indexing and mining one billion time series. In: IEEE International Conf. on Data Mining, pp. 58–67 (2010)
4. Chávez, E., Navarro, G., Baeza-Yates, R., Marroquín, J.L.: Searching in metric spaces. ACM Computing Surveys 33(3), 273–321 (2001)
5. Hetland, M.L.: Ptolemaic indexing. arXiv:0911.4384 [cs.DS] (2009)
6. Junkui, L., Yuanzhen, W.: Early abandon to accelerate exact dynamic time warping. Int. Arab. J. Inf. Technol. 6(2), 144–152 (2009)
7. Keogh, E.: Exact indexing of dynamic time warping. In: Proceedings of the 28th International Conference on Very Large Data Bases, pp. 406–417 (2002)
8. Keogh, E., Ratanamahatana, C.A.: Exact indexing of dynamic time warping. Knowledge and Information Systems 7, 358–386 (2005), doi:10.1007/s10115-004-0154-9
9. Keogh, E., Xi, X., Wei, L., Ratanamahatana, C.: The UCR Time Series Classification/Clustering Homepage (2006)
10. Kim, S.-W., Park, S., Chu, W.W.: An index-based approach for similarity search supporting time warping in large sequence databases. In: Proceedings of the 17th International Conference on Data Engineering, pp. 607–614. IEEE Computer Society, Washington, DC (2001)
11. Lokoč, J., Hetland, M.L., Skopal, T., Beecks, C.: Ptolemaic indexing of the signature quadratic form distance. In: Proceedings of the Fourth International Conference on Similarity Search and Applications, pp. 9–16. ACM (2011)
12. Mico, M.L., Oncina, J., Vidal, E.: A new version of the nearest-neighbour approximating and eliminating search algorithm (aesa) with linear preprocessing time and memory requirements. Pattern Recogn. Lett. 15(1), 9–17 (1994)
13. Rabiner, L., Juang, B.-H.: Fundamentals of speech recognition. Prentice-Hall, Inc., Upper Saddle River (1993)
14. Sakoe, H.: Dynamic programming algorithm optimization for spoken word recognition. IEEE Trans. on Acoustics, Speech, and Signal Processing 26, 43–49 (1978)
15. Shieh, J., Keogh, E.J.: iSAX: disk-aware mining and indexing of massive time series datasets. Data Min. Knowl. Discov. 19(1), 24–57 (2009)
16. Skopal, T.: On Fast Non-metric Similarity Search by Metric Access Methods. In: Ioannidis, Y., Scholl, M.H., Schmidt, J.W., Matthes, F., Hatzopoulos, M., Böhm, K., Kemper, A., Grust, T., Böhm, C. (eds.) EDBT 2006. LNCS, vol. 3896, pp. 718–736. Springer, Heidelberg (2006)
17. Skopal, T.: Unified framework for fast exact and approximate search in dissimilarity spaces. ACM Transactions on Database Systems 32(4), 1–46 (2007)
18. Yi, B.-K., Jagadish, H.V., Faloutsos, C.: Efficient retrieval of similar time sequences under time warping. In: Proceedings of the Fourteenth International Conference on Data Engineering, ICDE 1998, pp. 201–208 (1998)
19. Zezula, P., Amato, G., Dohnal, V., Batko, M.: Similarity Search: The Metric Space Approach (Advances in Database Systems). Advances in Database Systems. Springer-Verlag New York, Inc., Secaucus (2005)

# A Appendix

## A.1 Proof of Theorem 1

*Proof.* The validation of first two conditions is straightforward, while the last one we need to prove by contradiction. We take Eq. 7 and for contradiction let suppose

$$\delta_p(x, y + \epsilon) > \delta_p(x, y)$$
$$|x - (y + \epsilon)|^p > |x - y|^p$$
$$x - y - \epsilon > x - y \quad (x \geq y + \epsilon, x \geq y)$$
$$0 > \epsilon$$

The last step shows the contradiction we obtained as the definition assumed $\epsilon \geq 0$. Similarly, the same principle applies for Eq. 8.

Therefore, we can conclude that any function from $\delta_p$ family is *LB-compliant*. □

## A.2 Proof of Theorem 2

*Proof.* The steps correspond to the proof of lowerbounding by LB_Keogh [7]. Therefore we omit some steps and for details we refer the reader to this paper.

For contradiction, let suppose $LB\_Tight(Q, S, \delta, f) > GDTW(Q, S, \delta, \omega, f)$. Then

$$f\left(\sum_{i=1}^{n} \begin{cases} \delta(q_i, u_i) & \text{if } q_i > u_i \\ \delta(q_i, l_i) & \text{if } q_i < l_i \\ 0 & \text{otherwise} \end{cases}\right) > f\left(\sum_{k=1}^{X} \delta(w_k)\right)$$

The function $f$ is monotonic and it applies to both sides, so we can discard it. Because $n \leq X$ (from warping path definition; see Eq. 2), every term on the left-hand side (LHS) will match a unique term on the right-hand side (RHS), thus leaving $X - n$ terms unmatched. So

$$\sum_{k=1}^{X} \delta(w_k) = \sum_{k \in \text{matched}} \delta(w_k) + \sum_{k \in \text{unmatched}} \delta(w_k)$$

As we map every $i$-th term from LHS with exactly one of $i$-th terms on the RHS, we will take the item on RHS with the lowest value of $j$ (there might be several $j$ values for a single $i$). Other items $w_k$ remain in the unmatched part. If we ignore the unmatched summation and take only the one with matched items, we will have three types of relationships between matched terms and LHS to consider. They all will lead to the conclusion that all matched terms are larger than their corresponding counterparts.

1. $q_i > u_i$

   In this situation we will observe the relationship between $\delta(q_i, u_i)$ and $\delta(w_k)$, particularly $\delta(q_i, u_i)$ and $\delta(q_i, s_j)$ as defined according to Eq.1.

   As both time series Q, S have the same length, $n = m$, this means that

   $$j - \omega \leq i \leq j + \omega$$

which leads to limitation of $j$ value

$$i - \omega \leq j \leq i + \omega$$

and it gives us

$$s_j \in \{s_{i-\omega}, \ldots, s_{i+\omega}\}$$

As $u_i = max(s_{i-\omega}, \ldots, s_{i+\omega})$ is the definition of the upper envelope, we will certainly have $u_i \geq s_j$. Moreover, we define $\epsilon = u_i - s_j \geq 0$. To continue with relationship checking, we get

$$\delta(q_i, u_i) = \delta(q_i, s_j + \epsilon) \leq \delta(q_i, s_j)$$

because the distance function $\delta$ is *LB-compliant* and we apply Eq. 7.

2. $q_i < l_i$

This case is handled similarly as the previous one, so we include only distinct steps. We start with the relationship between $\delta(q_i, l_i)$ and $\delta(w_k) = \delta(q_i, s_j)$. With $l_i = min(s_{i-\omega}, \ldots, s_{i+\omega})$ as the definition of the lower envelope, we will have $l_i \leq s_j$. Now, we define $\epsilon = s_j - l_i \geq 0$. This means that

$$\delta(q_i, s_j) = \delta(q_i, l_i + \epsilon) = \delta(l_i + \epsilon, q_i) \geq \delta(l_i, q_i)$$

because the distance function $\delta$ is *LB-compliant* and we used the third condition (see Eq. 8). Additionally, the $\delta$ function must be symmetric, so we get $\delta(q_i, s_j) \geq \delta(q_i, l_i)$

3. The last case confirms that $0 \leq \delta(q_i, s_j)$. This results directly from the definition of function $\delta$ which must be non-negative (the first condition of *LB-compliant* functions).

The above mentioned cases revealed that all matched terms in $\sum_{k \in \text{matched}} \delta(w_k)$ are larger than their matching counterparts on LHS. As the distance function $\delta$ is non-negative, the remaining summation $\sum_{k \in \text{unmatched}} \delta(w_k)$ cannot be negative. So, we reached a contradiction as the summations on RHS are greater or equal than the summation on LHS. Thus our assumption was incorrect, so the previous statement holds:

$$\texttt{LB\_Tight}(Q, O, \delta, f) \leq GDTW(Q, O, \delta, \omega, f)$$

$\square$

# A Multivariate Correlation Distance
# for Vector Spaces

Richard Connor and Robert Moss

Department of Computer and Information Sciences,
University of Strathclyde, Glasgow G1 1XH
Scotland, UK
{richard.connor,robert.moss}@strath.ac.uk

**Abstract.** We investigate a distance metric, previously defined for the measurement of structured data, in the more general context of vector spaces. The metric has a basis in information theory and assesses the distance between two vectors in terms of their relative information content. The resulting metric gives an outcome based on the dimensional correlation, rather than magnitude, of the input vectors, in a manner similar to Cosine Distance.

In this paper the metric is defined, and assessed, in comparison with Cosine Distance, for its major properties: semantics, properties for use within similarity search, and evaluation efficiency.

We find that it is fairly well correlated with Cosine Distance in dense spaces, but its semantics are in some cases preferable. In a sparse space, it significantly outperforms Cosine Distance over *TREC* data and queries, the only large collection for which we have a human-ratified ground truth. This result is backed up by another experiment over *movielens* data. In dense Cartesian spaces it has better properties for use with similarity indices than either Cosine or Euclidean Distance. In its definitional form it is very expensive to evaluate for high-dimensional sparse vectors; to counter this, we show an algebraic rewrite which allows its evaluation to be performed more efficiently.

Overall, when a multivariate correlation metric is required over positive vectors, SED seems to be a better choice than Cosine Distance in many circumstances.

**Keywords:** distance metric, multivariate correlation, vector space, cosine distance, similarity search.

## 1 Introduction

Structural Entropic Distance (SED) was developed as a distance metric over unordered tree structures [1], and has been extended to handle a number of other types of structured data [3]. The essence of the metric is to use a structured object as a conceptual generator of events which are then represented in the form of event *ensembles*, the type of event/probability map used by Shannon to calculate entropy [12].

G. Navarro and V. Pestov (Eds.): SISAP 2012, LNCS 7404, pp. 209–225, 2012.
© Springer-Verlag Berlin Heidelberg 2012

In this paper, we consider the same function redefined as a vector distance metric; in fact more accurately as a class of metrics over $\mathbb{R}^n$ for any $n$. For any set of ensembles, an isomorphism to a set of vectors exists where each event is assigned a unique number, and the probability of that event is assigned to the scalar component of that dimension of the vector. Zeros are assigned for events that do not occur. For example:

$$E_A = \{a : 0.3, b : 0.1, c : 0.6\} \quad \Rightarrow \quad V_A \; = [0.3, 0.1, 0.6, 0, 0]$$
$$E_B = \{a : 0.8, d : 0.6\} \quad \Rightarrow \quad V_B \; = [0.8, 0, 0, 0.6, 0]$$
$$E_C = \{b : 9, e : 1\} \quad \Rightarrow \quad V_C \; = [0, 9, 0, 0, 1]$$

Using such an isomorphism, it is straightforward to formulate a distance metric $D_V$ over any positive vector space such that

$$D_V(V_i, V_j) = D_E(E_i, E_j)$$

where $D_E$ is the structural distance defined over ensembles in [2]. $D_V$ is formally defined in Section 3 of this paper.

SED-vector is a distance metric suitable for magnitude-independent multivariate correlation; the only other proper metric that we know in this class is a variant of Cosine Distance. In this paper we show properties of the metric with respect to its semantics, its indexing properties, and its efficiency, as follows:

**Semantics:** Over dense spaces, it has a fairly tight correlation with Cosine distance, but some different properties that may be more desirable.
Over sparse spaces, we explain a significant improvement over Cosine distance, evidenced by experiment with real-world data.

**Indexing:** Over dense spaces, SED-vector has lower intrinsic dimensionality and better near-origin density than either Cosine or Euclidean distances.

**Cost:** The metric as defined appears to demand a costly complexity evaluation to be made for each new comparison. We show an algebraic rewrite of the original formula which allows a much more efficient evaluation.

Our conclusion is that where the semantics of a data collection dictate that component correlation is more important than magnitude, SED will usually be a better choice than Cosine distance for the application of similarity search.

## 2    Multivariate Correlation Distance

The vector is a convenient data structure to store multivariate data. For example, the MPEG-7 standard [7] defines a set of scalar characteristics which can be deduced from an image, including numeric classification of values such as colour histograms and edge features. Each character is assigned a unique dimension in a multi-dimensional vector space, and vectors used to represent the images. Similarity search over images can then proceed using these vectors as proxies.

In many cases, the relative ratios of scalar values within dimensions is more significant than the absolute magnitude of the scalar components. For example,

during a currency crisis, every stock in a stock exchange may decrease in value
by exactly the same ratio; when looking for similarities in trading positions, a
vector representing stocks which have all decreased by the same ratio may be
more similar than one where most of the scalar values are the same, but a few are
relatively different. During the gathering of insects in a biodiversity investigation,
the absolute numbers gathered may depend on constant, external factors and
again the relative ratios among species are likely to be more important. The
best-known example where ratios are generally judged to be more important
than absolutes is in Information Retrieval, where documents of arbitrary size
are compared against each other by their term frequencies, rather than term
numbers [13].

There is surprisingly little work in the consideration of distance metrics in
this context. A bit vector form of Jaccard distance was noted by Tanimoto[1] in
[11], where the Jaccard set distance is written as

$$D_T(v, w) = 1 - \frac{v \cdot w}{|v|^2 + |w|^2 - v \cdot w}$$

where $v$ and $w$ are bit vectors representing sets. There is a great deal of confusion
over this function, which is regularly cited as a proper distance metric over
vectors; however, it is not. It is a proper metric over sets, and over weighted sets,
when they are represented in vector form. [8] proves this property for weighted
sets only, and not vectors in general. In the context of weighted sets, it is very
commonly used as a proxy for molecule similarity by organic chemists.

Cosine distance has been used since the birth of Information Retrieval [9] [10]
for this purpose, and perhaps for this reason is almost the *de facto* multivariate
correlation function. However there is also significant confusion over this. Cosine
distance is most commonly defined as the convenient algebraic form:

$$D_C(v, w) = 1 - \frac{v \cdot w}{|v||w|}$$

The key observation is that the outcome of this is independent of the magnitude
of the vectors, and is within the interval [0,1], giving zero for perfectly aligned
vectors and one for independent vectors[2].

There are two problems, however. The first is that this is not a proper metric
as it does not have triangle inequality. Probably as a result of this, the term
"Cosine Distance" is also used to mean simply the angle between the vectors,
normalised into the range $[0, 1]$ by division by $\frac{\pi}{2}$ for positive vectors, e.g. [4]. Both
functions give the same ordering over any set of comparisons, but the second also
gives triangle inequality. We therefore use the term "Cosine Distance" to mean

$$D_{Cos}(v, w) = \frac{2 \cdot \cos^{-1}\left(\frac{v \cdot w}{|v||w|}\right)}{\pi}$$

from this point onwards.

---

[1] In fact, Tanimoto neither defined nor used this as a distance function, but was
interested in the function $-log_2(D_T(A, B))$.

[2] Those with no common non-zero dimensions.

The second problem, as will be elaborated, is that vector alignment does not serve as a good proxy for multivariate correlation in a number of circumstances. Singhal [13] states: "Typically, the angle between two vectors is used as a measure of divergence between the vectors, and cosine of the angle is used as the numeric similarity (since cosine has the nice property that it is 1.0 for identical vectors and 0.0 for orthogonal vectors)." It would appear that cosine similarity has been in use in this field for 50 years without much more attention than this being paid to its efficacy.

We therefore introduce SED-vector as a competitor function which we claim has, in general, better properties for similarity search.

## 3  Definition

We restrict the text here to a succinct definition of the SED metric within the context of vectors and a short statement of intuition; [2] gives a full description of its derivation and properties.

### 3.1  Domain Definitions

The domain is a set $V_l$ of vectors where $l$ denotes a fixed (arbitrary) number of dimensions, with vectors $v, w$ etc. drawn from $V_l$. The scalar component of the $i$th dimension of a vector $v$ is denoted $v_i$. The domain is restricted to positive vectors, i.e. $\forall v \in V_l, \ \forall i \leq l, \ v_i \geq 0$.

### 3.2  Definition of SED over Vectors

The *Structural Entropic Distance (SED)* between two vectors is defined in terms of a complexity function over vectors, based on their information content. For this purpose we consider the normalised form of a vector $v$ as $\overline{v}$, which is scaled so that $\overline{v} = \alpha v$ for some $\alpha \in \mathbb{R}$, and $\sum_i \overline{v}_i = 1$. The information content of a vector is defined using Shannon's entropy function:

$$H_x(v) = -\sum_i \overline{v}_i \log_x(\overline{v}_i)$$

noting that $H_x(\overline{v}) = H_x(v)$ in all cases.

The complexity function $C$ removes the dependence on the logarithm base:

$$C(v) = x^{H_x(v)}$$

and the distance is defined in terms of the complexity of the two vectors, and the complexity of the sum of the (normalised) vectors:

$$D(v, w) = \frac{C(\overline{v} + \overline{w})}{\sqrt{C(v) \cdot C(w)}} - 1 \tag{1}$$

## 3.3   Underlying Intuition

The function is based upon the comparison of the information content of the two vectors under comparison, and a third vector which represents an information-theoretic "merge". The central term of Equation (1) gives the complexity ratio of the "merge" to the geometric mean of its components. If the vectors are identical, then this ratio is 1; if they are independent, then the ratio is 2.

By corollary, a vector formed through the merging of two others which have no non-zero dimensions in common, and each of which has the same complexity, contains twice the amount of information as each individual vector, which is a reasonable intuition. Given our definition of complexity, that intuition holds only when the complexity of each vector is the same, the geometric mean being used to balance these terms when they are different.

## 4   Metric Properties

Proofs of the metric properties of this function defined ensembles are given in [2]; most of these can be reapplied to the more general vector space considered here, and are not repeated.

- SED is positive and symmetric.
- It is bounded in the interval $[0, 1]$. It returns 1 if and only if its input comprises two *independent* vectors, i.e. they have no non-zero terms in common. It returns 0 if and only if its input comprises two vectors one of which is a scalar multiple of the other. It thus has the pseudo-identity property.
- SED as defined so far is not a proper distance metric in generalised vector space. There exist vector triples over which $D$ does not preserve triangle inequality, the simplest and worst case being the vectors $[0, 1]$, $[1, 0]$ and $[1,1]$, as $D([0, 1], [1, 1]) = 0.24$ and $D([0, 1], [1, 0]) = 1.0$.
  $D$ has previously been used as a distance metric over restricted subsets of the vector domain where cases such as this do not occur. To achieve triangle inequality in generalised vector space the requirement is to raise the numeric outcome of $D$ to the power 0.48.[3]

## 5   Semantic Properties

### 5.1   Analysis

**Comparison of SED-Vector and Cosine Distance.** SED-vector and Cosine distances are highly correlated over many data sets. However there are two important circumstances where their outcomes of the two metrics are significantly different:

**Small Value Pairs:** when there is a large relative difference in the magnitude of one or more scalar component pairs, SED assigns much more significance than Cosine to the differences between the smaller components; and

---

[3] A conservative value of $x$ where $0.24^x \geq 0.5$; see [2] for details.

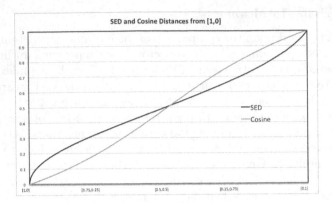

**Fig. 1.** SED and Cosine Distance over Unit Vectors. The x-axis represents vectors of the form $[1 - x, x]$ for values of $x$ ranging from 0.0 to 1.0; the y-axis plots the distance between these and the vector $[1,0]$.

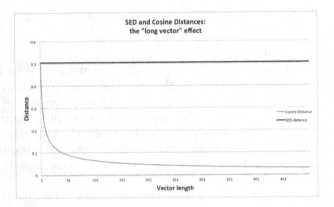

**Fig. 2.** SED and Cosine Distance over increasing length vectors. The graph shows the distance between vectors of the form $[1,0,...,0]$ and $[1, \frac{1}{n-1}, ..., \frac{1}{n-1}]$, where $n$ is the length of the vector. The distance is plotted against the number of independent dimensions. These pairs each have the same information overlap, but with the independent terms spread over increasingly many dimensions.

**Sparse Vectors:** when a significant proportion of one or both vectors is independent, i.e. many dimensions are positive in one vector and zero (or very small) in the other, then the outcome of Cosine distance is significantly affected by the distribution of the values of these dimensions, whereas SED is unaffected.

Figure 1 shows the vector $[1,0]$ compared by both metrics against all other two-dimensional vectors whose components sum to 1, from $[1,0]$ to $[0,1]$. Both metrics give distances starting at 0 for the identical vectors, increasing monotonically to 1 for the distance between $[1,0]$ and $[0,1]$. However the notable difference between the two functions is how quickly SED increases for very small values in the first dimension, illustrating the first difference highlighted above.

Which behaviour is more desirable depends deeply on the meaning of the vectors under consideration. SED will be much more effective at highlighting large relative differences in small values, which in Cosine distance will be masked by much smaller relative differences in larger values.

Figure 2 shows an effect which, in terms of information overlap, shows an objectively better behaviour for SED over Cosine. The graph plots distances for a series of successively longer pairs of vectors, each with a 1 in the first dimension, but where the first is padded out with zeros, and the second is padded out with $\frac{1}{n-1}$ where n is the length of the vector. The point of the experiment is that each pair of vectors has exactly the same information overlap, and the distribution of the residue should not be significant in information-theoretic terms.

In this case it can be seen that Cosine distance drops quite sharply according to this increase in dimension. This behaviour gives false results in these cases, when the metric is being used as a measure of correlation - there is no intuitive reason why an increase in the number of non-correlated dimensions should decrease the distance. As shown algebraically in Section 7, SED has the property that the distribution of independent dimensions does not affect the outcome.

Both of these differences are a direct result of Cosine distance using vector alignment as a proxy for correlation. We are not aware of any discussion on the origins of this use, other than a purely pragmatic observation that the use of vector alignment as a proxy for similarity removes any significance from the magnitude of the vectors under comparison.

The first effect is intuitively fairly clear: when one dimension of a vector is orders of magnitude larger than others, then the effect of the smaller dimensions on the alignment of the vector is negligible. The increase in the information content of an event stream caused by a new low frequency event is much greater.

The second effect is also clear as a consequence of vector geometry; as a single scalar component is spread over multiple dimensions, the effect on the alignment of the vector reduces until it is insignificant. This effect is probably never an intention for Cosine distance, and in fact is documented as a problem in the domain of Information Retrieval[4] [14].

## 5.2 Experimental Evaluation

To establish that our theoretically-based observations are seen in practice, we have performed experiments to measure SED against Cosine for a number of data sets:

1. To observe the "small value" property, we compare the two metrics over randomly generated vectors within a five-dimensional Cartesian space.

---

[4] In the TF-IDF model, longer documents tend to have more terms each with lower frequency, and Cosine distance gives smaller outcomes for longer documents with the same ratio of search terms.

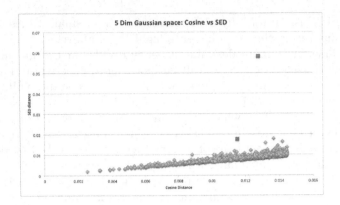

**Fig. 3.** Correlation of SED against Cosine Distance

2. To show the "sparse vector" property, we show how the two metrics compare against two real-world data collections: *TREC* [5] and *movielens* [15]. The outcome over *TREC* is particularly significant, as it is the only large data collection for which we have a ratified ground truth available. We show that SED hugely outperforms Cosine distance.

**5-Dimensional Cartesian Space.** To establish the correlation and differences between SED and Cosine Distance over dense vector spaces, we performed[5] correlation analysis over a set of $10^9$ distances evaluated from the comparison of 14,143 randomly-generated 5-dimensional Cartesian vectors, each dimension being generated within a Gaussian distribution with $\mu = 0.5$ and $\sigma = 0.2$, restricted to within [0,1]. As in this context we are interested particularly in small distances that are likely to lie within a query threshold, we restrict our presentation to the smallest $10^{-6}$ of distances observed for either metric.

Figure 3 shows a scatter-plot of SED against Cosine distance, while Figure 4 shows Cosine distance against SED. The general correlation is quite apparent, and in particular it can be seen by comparing the two graphs that around 90% of the same data points occur in both. It can also be seen that a consistent pattern of outliers exists, which are points showing a (relative) closeness under Cosine distance where that closeness is not reflected by SED.

From the above analysis, our hypothesis is that the outliers are likely to be points where one or more of the component pairs are relatively small values which are significantly different from each other. For the outlier distances highlighted in both graphs, the underlying vectors pairs are shown in Tables 1a and 1b respectively, the tables also including from each a "normal" point drawn from the same x-coordinate as the worst outlier. From inspection of these and other points it can be seen that this is indeed the reason why some points are deemed to be relatively closer using Cosine distance. This effect turns out to underlie, for each metric, the 10% of points not included by the other.

---

[5] The Java source code for this and all other experiments is available from the authors.

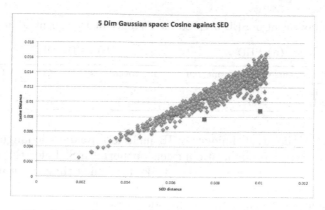

**Fig. 4.** Correlation of Cosine Distance against SED

Due to the use of Gaussian distributions within the vector components, these points of disagreement are quite rare. When a flat distribution is used over 5 dimensions, the same effects are seen, but the percentage of points included by both metrics drops to around 70% and the number of outliers increases correspondingly.

**Table 1.** Outlier vector pairs highlighted

**(a)** Outlier vector pairs highlighted in Figure 3

| Outlier | Vector 1 | Vector 2 | Distance Cosine | SED |
|---|---|---|---|---|
| Outlier 1 | $\begin{pmatrix}0.512\\0.44\\0.429\\0.696\\0.016\end{pmatrix}$ | $\begin{pmatrix}0.464\\0.401\\0.404\\0.633\\0.0\end{pmatrix}$ | 0.013 | 0.058 |
| Outlier 2 | $\begin{pmatrix}0.034\\0.434\\0.258\\0.477\\0.572\end{pmatrix}$ | $\begin{pmatrix}0.063\\0.619\\0.382\\0.689\\0.807\end{pmatrix}$ | 0.0115 | 0.0175 |
| "Normal" | $\begin{pmatrix}0.449\\0.499\\0.3\\0.485\\0.733\end{pmatrix}$ | $\begin{pmatrix}0.392\\0.433\\0.259\\0.426\\0.662\end{pmatrix}$ | 0.0115 | 0.0075 |

**(b)** Outlier vector pairs highlighted in Figure 4

| Outlier | Vector 1 | Vector 2 | Distance Cosine | SED |
|---|---|---|---|---|
| Outlier 1 | $\begin{pmatrix}0.14\\0.474\\0.562\\0.64\\0.678\end{pmatrix}$ | $\begin{pmatrix}0.137\\0.43\\0.523\\0.588\\0.617\end{pmatrix}$ | 0.0075 | 0.0077 |
| Outlier 2 | $\begin{pmatrix}0.438\\0.426\\0.1\\0.371\\0.587\end{pmatrix}$ | $\begin{pmatrix}0.429\\0.423\\0.109\\0.373\\0.578\end{pmatrix}$ | 0.0088 | 0.01 |
| "Normal" | $\begin{pmatrix}0.415\\0.591\\0.293\\0.476\\0.589\end{pmatrix}$ | $\begin{pmatrix}0.423\\0.618\\0.32\\0.505\\0.639\end{pmatrix}$ | 0.0101 | 0.0142 |

***TREC* and TF/IDF.** In Information Retrieval (IR), documents may be treated as "bags of words" , represented by sparse vectors where each dimension corresponds to an individual term in the lexicon. The number of occurrences of a term within a document is given by the corresponding dimension of the vector.

**Table 2.** Improvement of SED over Cosine distance at Document 20 in rank order

<table>
<tr><td colspan="3" align="center">(a) SED</td><td colspan="3" align="center">(b) SED/IDF</td></tr>
<tr><td>Improvements</td><td>Cosine</td><td>Cosine/IDF</td><td>Improvements</td><td>Cosine</td><td>Cosine/IDF</td></tr>
<tr><td>Precision</td><td>109%</td><td>28%</td><td>Precision</td><td>157%</td><td>57%</td></tr>
<tr><td>Recall</td><td>218%</td><td>15%</td><td>Recall</td><td>282%</td><td>38%</td></tr>
</table>

Other state of the art techniques in IR include probabilistic language models such as BM25. While probabilistic models outperform SED for the simple task of document retrieval, there are other tasks for which they make no sense, or for which vector space models remain the dominant technique; for example in pseudo-relevance feedback and recommender systems, which try to make predictions based on document similarity.

We use the Congressional Record of the 103rd Congress (1993 - Text Research Collection Volume 4) and the Los Angeles Times (1989, 1990 - Text Research Collection Volume 5) from the Text REtrieval Conference (*TREC*) series test collections. This comprises: a set of documents; a set of topics, and a corresponding set of relevance judgments – a total of $500k$ documents, a vocabulary size of $525k$ words, and $170M$ word occurrences.

Topics 301-350 and 351-400 from the *TREC* 6 and 7 'ad hoc' collections respectively, were used to compare Cosine distance with SED. For the 100 queries, the distance between each document in the corpus and the query were taken. The 20 nearest documents for each query were selected and compared to the provided relevance judgements to determine whether it was a match. Relevance is a judgement ratified by human readers [5].

As each returned document is assessed for relevance, two running scores are kept: precision – the number of correctly identified relevant documents divided by the number of documents assessed; and recall – the number of correctly identified relevant documents divided by the number of documents that should have been identified as relevant. For every query, the precision and recall were calculated for each of the 20 ranks, then at each rank the mean precision and recall across all queries for that rank were plotted.

SED has better precision and recall for every rank than Cosine distance, as shown in Figure 5 and Table 2. Perhaps the most remarkable outcome is that SED without IDF significantly outperforms Cosine with IDF; we believe this is because of SED's better performance over the independent dimensions of the very large and sparse vectors.

**User Rankings for Movies.** A further semantic test was performed by experiment over the *movielens* data collection [15].

This collection comprises a set of real users' movie rankings. We interpret the data as a collection of sparse vectors, with one vector for each user, and each dimension corresponding to a particular movie. Vector values are ratings given by users to movies on a scale of 1 to 5, with a zero indicating that the movie has not been rated.

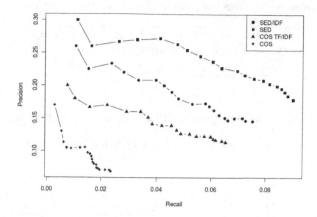

**Fig. 5.** *TREC* precision/recall diagram. In general, the quality of the retrieval technique is correlated with the area under the graph. SED can be seen to greatly outperform Cosine distance.

To use this collection for a semantic test, we use the underlying assumption that significant correlations exist between like-minded users - this hypothesis has been adequately tested by successful research on recommender systems [6]. We partitioned the data into two collections of vectors, indexed by the same user identities in both collections, with each collection containing half of the original movies. For each user, we then used each distance function to find the closest user within the first set, and compared this distance with the distance between the same pair of users in the second set; for a perfect semantic distance, and a perfect data set, these distances should be the same. We therefore used Pearson's product-moment coefficient over all the distance pairs obtained to give a score to the distance metric.

The 1M data set was used, comprising just over one million ratings by six thousand users for four thousand movies. To allow repetition over independent data sets, the users were split into six different groups of a thousand users each; each collection therefore contained one thousand sparse vectors, each of two thousand dimensions. The correlations obtained for each pair of files, for each distance metric, are shown in Table 3.

It can be seen that in all these cases, SED gave a better outcome than Cosine Distance. The mean difference in outcome is 0.036, and the standard deviation of the difference is 0.0088, making the mean just over 4 standard deviations above zero. Using the null hypothesis that the choice of metric makes no difference to the correlation of the outcome, this gives a $p_0$ value of $10^{-5}$, strongly justifying the hypothesis that SED is a better semantic similarity function.

**Table 3.** Pearson correlations for pairs of distances taken from *movielens* data

| Metric | test1 | test2 | test3 | test4 | test5 | test6 | mean |
|--------|-------|-------|-------|-------|-------|-------|------|
| SED    | 0.835 | 0.859 | 0.837 | 0.852 | 0.901 | 0.849 | 0.855 |
| Cosine | 0.784 | 0.823 | 0.796 | 0.827 | 0.876 | 0.810 | 0.819 |

**Table 4.** Key values for the Histograms of Figure 6, and for 10-dim Gaussian vectors

**(a)** 5-dim Gaussian

| Metric | $\mu$ | $\sigma$ | IDIM | $\mu$-point |
|--------|-------|----------|------|-------------|
| SED       | 0.20 | 0.08 | 3.0 | 0.006 |
| Cosine    | 0.29 | 0.10 | 3.7 | 0.008 |
| Euclidean | 0.27 | 0.08 | 5.0 | 0.014 |

**(b)** 5-dim Non-Gaussian

| Metric | $\mu$ | $\sigma$ | IDIM | $\mu$-point |
|--------|-------|----------|------|-------------|
| SED       | 0.31 | 0.11 | 4.0 | 0.009 |
| Cosine    | 0.42 | 0.14 | 4.3 | 0.011 |
| Euclidean | 0.39 | 0.11 | 6.3 | 0.021 |

**(c)** 10-dim Gaussian

| Metric | $\mu$ | $\sigma$ | IDIM | $\mu$-point |
|--------|-------|----------|------|-------------|
| SED       | 0.21 | 0.06 | 6.8  | 0.034 |
| Cosine    | 0.31 | 0.08 | 8.7  | 0.051 |
| Euclidean | 0.27 | 0.06 | 10.3 | 0.052 |

# 6 Indexing Properties

Figure 6 shows histograms of SED, Cosine and Euclidean distances over randomly generated 5-dimensional vectors, with both Gaussian and non-Gaussian internal distributions. The key values summarising these are displayed in Table 4, giving values for median, standard deviation, intrinsic dimensionality (*IDIM*), and the upper bound of the smallest $10^{-6}$ of the distances measured ($\mu$-point).

The latter two values are key for performance within similarity search. *IDIM* has been shown to be a good predictor of performance for a given query threshold, and the $\mu$-point is an estimator useful for determining query thresholds. In general, the lower these values, the better will be the performance of a similarity search index using the metric.

As can be seen from Table 4, SED gives better properties for use in metric indices than Cosine distance. We have observed this to be true over a range of different dimensions for dense spaces such as these. The comparison with Euclidean space is included to give an idea of how many dimensions might be tractable for indexing purposes for strategies where this is already known for Euclidean distance.

# 7 Cost of Evaluation

The definition of *SED* as presented in terms of complexity requires significant computation. In this section it is shown via algebraic manipulation how an alternative formulation with a lower computational cost can be achieved. The term:

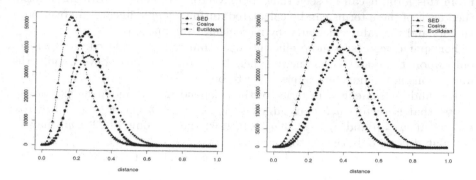

**Fig. 6.** 5-dimensional Histograms. The left-hand graphs shows histograms over 5-dimensional Cartesian vectors each dimension of which has a Gaussian distribution, the right-hand graphs shows 5-dim vectors each dimension of which has a flat distribution. Euclidean distance is normalised into the interval [0,1].

$$\frac{C(\overline{v} + \overline{w})}{\sqrt{C(v) \cdot C(w)}}$$

is defined as

$$\frac{x^{H_x(\overline{v}+\overline{w})}}{\sqrt{x^{H_x(v)} \cdot x^{H_x(w)}}}$$

and so can be rewritten as

$$x^{H_x(\overline{v}+\overline{w}) - \frac{1}{2}(H_x(v) + H_x(w))}$$

Substituting over the definition of $H_x$, this exponent of $x$, for any log base $x$, can be written as[6]

$$\sum_i \left(\frac{\overline{v}_i + \overline{w}_i}{2}\right) \overline{\log}_x \left(\frac{\overline{v}_i + \overline{w}_i}{2}\right) -$$

$$\frac{\sum_i \overline{v}_i \overline{\log}_x(\overline{v}_i) + \sum_i \overline{w}_i \overline{\log}_x(\overline{w}_i)}{2}$$

or as a single summation:

$$\frac{1}{2} \sum_i \left((\overline{v}_i + \overline{w}_i)\overline{\log}_x\left(\frac{\overline{v}_i + \overline{w}_i}{2}\right) - \right.$$

$$\left. \overline{v}_i \overline{\log}_x(\overline{v}_i) - \overline{w}_i \overline{\log}_x(\overline{w}_i)\right) \qquad (2)$$

---

[6] To make the arithmetic clearer, negative expressions are rewritten using the notation $\overline{\log}(x)$ to denote $-\log(x)$; as arguments are less than one, all such terms are positive.

From this form, it can be seen that, if the vectors are kept in normalised form, then each of these terms can be calculated during a single linear scan over both vectors and the value calculated by incremental summation.

For sparse vectors, a more efficient algorithm can be achieved, which relies only upon the scalar components when both are non-zero allowing only the dimensions in the non-zero intersection to be accessed.

If $\overline{v}$ and $\overline{w}$ are defined in terms of their dependent and independent components, that is vectors $a, b, c$ and $d$ such that $\overline{v} = a + b$ and $\overline{w} = c + d$, where $b_i = \overline{v}_i$ iff $w_i = 0$ and $d_i = \overline{w}_i$ iff $v_i = 0$, then the outcome of $D(v, w)$ may be calculated using only the vectors $a$ and $c$.

The term (2) can be written in terms of the component vectors as

$$\frac{1}{2}\sum_i \left( \left(a_i + c_i\right)\overline{\log}_x\left(\frac{a_i + c_i}{2}\right) - a_i\overline{\log}_x(a_i) - c_i\overline{\log}_x(c_i) \right)$$

$$+$$

$$\frac{1}{2}\sum_i \left( b_i\,\overline{\log}_x\left(\frac{b_i}{2}\right) - b_i\overline{\log}_x(b_i) \right)$$

$$+$$

$$\frac{1}{2}\sum_i \left( d_i\,\overline{\log}_x\left(\frac{d_i}{2}\right) - d_i\overline{\log}_x(d_i) \right)$$

by noting that only one of these terms can be non-zero for each value of $i$. Now:

$$\frac{1}{2}\sum_i \left( b_i\,\overline{\log}_x\left(\frac{b_i}{2}\right) - b_i\overline{\log}_x(b_i) \right)$$

can be rewritten if we choose to use base 2 logarithms as:

$$\frac{1}{2}\sum_i b_i$$

The entire calculation can then be represented in terms of the component vectors, *only* for base 2 logarithms, as

$$\frac{1}{2}\sum_i \left( (a_i + c_i)\overline{\log}_2\left(\frac{a_i + c_i}{2}\right) - a_i\,\overline{\log}_2(a_i) - c_i\,\overline{\log}_2(c_i) + b_i + d_i \right) \tag{3}$$

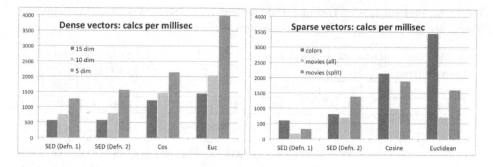

**Fig. 7.** Performance of two versions of SED, compared with Cosine and Euclidean distances over the same data sets. A significant speedup according to the algebraic rewrite (SED Defn. 2) can be observed over sparse vectors with small intersections.

Now noting that $\sum_i b_i = 1 - \sum_i a_i$, and similarly for $c$ and $d$:

$$\frac{1}{2} \sum_i \left( (a_i + c_i) \overline{\log}_2 \left( \frac{a_i + c_i}{2} \right) - a_i \overline{\log}_2(a_i) - c_i \overline{\log}_2(c_i) \right)$$

$$+ \left( \frac{1}{2} - \frac{1}{2} \sum_i a_i \right) + \left( \frac{1}{2} - \frac{1}{2} \sum_i c_i \right)$$

which can itself be written as

$$1 - \frac{1}{2} \sum_i \left( (a_i \overline{\log}_2(a_i) + c_i \overline{\log}_2(c_i)) \right.$$

$$\left. - (a_i + c_i) \overline{\log}_2 \left( \frac{a_i + c_i}{2} \right) + a_i + c_i \right)$$

demonstrating how the calculation can be performed with regard only to the dependent dimensions of the vector.

Figure 7 shows the resulting increase in efficiency, over sparse vectors in particular, and a general performance comparison with Cosine and Euclidean distance calculations over some representative data sets.

The left-hand chart shows performance over randomly-generated vectors, the right-hand chart shows performance over sparse vectors taken from the SISAP *colors* data set [4], and two sets of vectors taken from the *movielens* 100k data set detailed above. The details of these sparse data sets are given in Table 5; it can be seen how both SED and Cosine start to outperform Euclidean distance as the non-zero intersections become smaller. While SED is intrinsically a more costly calculation than either Cosine or Euclidean distance, it can be seen that the difference can be kept to within a small factor, and it may even be faster than Euclidean distance for sparse vectors with small intersections.

**Table 5.** Performance of Distance Calculations over Sparse Vectors. (Performance figures are given as the number of calculations performed per millisecond.)

| Collection | Dimensions | Mean non-zero | Mean intersection | SED | Cos. | Euc.n |
|---|---|---|---|---|---|---|
| movies (split) | 800 | 53 | 9 | 1386 | 1900 | 1602 |
| movies (all) | 1600 | 106 | 19 | 704 | 1004 | 709 |
| colors | 120 | 40 | 31 | 810 | 2142 | 3463 |

Dense vectors were represented by arrays of doubles, sparse vectors by two arrays, one of integers containing the non-zero indices, and one containing the data. For all metrics, the calculations were optimised as far as possible by using data objects to contain these arrays along with cached values for all calculations performed once per vector, such as complexity, magnitude, and component log values. Performance was measured by Java implementations of the metrics running on a 1.8 GHz Intel Core i7 with 4 GB 1333 MHz DDR3, enough to contain the working set for the measurements. The time for one million randomly selected distance calculations was repeatedly measured on an otherwise idle machine until the standard error of the mean was less than 0.5%, and median values reported.

## 8    Conclusions

We have presented an investigation into a previously reported distance metric, SED, applied as a multivariate correlation distance over generalised vector spaces. We have shown that in terms of both semantics and indexing properties, it is often likely to be a better alternative than Cosine Distance, while its cost has been shown to remain tractable in comparison. As Cosine Distance is generally believed to be the state-of-the-art for such data, we believe this is a significant contribution.

**Acknowledgements.** We would particularly like to thank an anonymous referee from a previous SISAP who questioned the behaviour of this function over vectors in general; now we know the answers to many of those questions. The referees from SISAP 2012 were also very helpful, in particular thanks for the suggestion of caching log values for the distance calculations which we should have thought of ourselves!

Morgan Harvey, an IR expert from The Friedrich-Alexander University, Erlangen-Nürnberg guided us in our analysis of SED vs Cosine distance over the *TREC* data collection and produced the precision/feedback chart presented here.

We would also like to acknowledge financial support from the UK Research Councils (EPSRC EP/G012407) and EU FP7 (D4SCIENCE, 212488 and D4S-II, 239019) without which this work would not have been possible.

# References

1. Connor, R., Simeoni, F., Iakovos, M.: Structural entropic difference: A bounded distance metric for unordered trees. In: International Workshop on Similarity Search and Applications, pp. 21–29 (2009)
2. Connor, R., Simeoni, F., Iakovos, M., Moss, R.: A bounded distance metric for comparing tree structure. Information Systems 36(4), 748–764 (2011)
3. Connor, R., Simeoni, F., Iakovos, M., Moss, R.: Towards a Universal Information Distance for Structured Data. In: SISAP 2011, Lipari, Italy, June 30-July 01 (2011)
4. Figueroa, K., Navarro, G., Chávez, E.: Metric Spaces Library, http://www.sisap.org/library/manual.pdf
5. Harman, D.K.: Overview of the rst Text REtrieval Conference (TREC-1). In: Proceedings of the First Text REtrieval Conference (TREC-1), 120 p. NIST Special Publication 500-207 (March 1993)
6. Herlocker, J., Konstan, J., Borchers, A., Riedl, J.: An Algorithmic Framework for Performing Collaborative Filtering. In: Proceedings of the 1999 Conference on Research and Development in Information Retrieval (August 1999)
7. ISO/IEC JTC 1/SC 29 (2009-10-30), MPEG-7 (Multimedia content description interface)
8. Lipkus, A.: A proof of the triangle inequality for the Tanimoto distance. Journal of Mathematical Chemistry 26(1), 263–265 (1999)
9. Maron, M.E.: An Historical Note on the Origins of Probabilistic Indexing. Information Processing and Management 44(2), 971–972 (2008), doi:10.1016/j.ipm.2007.02.012.
10. Maron, M.E., Kuhns, J.: On relevance, probabilistic indexing and information retrieval. Journal of the Association for Computing Machinery 7(3), 216–244 (1960)
11. Rogers, D.J., Tanimoto, T.T.: A Computer Program for Classifying Plants. Science 132 (October 1960)
12. Shannon, C.E.: A mathematical theory of communication. SIGMOBILE Mob. Comput. Commun. Rev. 5, 3–55 (2001)
13. Singhal, A.: Modern Information Retrieval: A Brief Overview. Bulletin of the IEEE Computer Society Technical Committee on Data Engineering 24(4), 35–43 (2001)
14. Singhal, A., Buckley, C., Mitra, M.: Pivoted document length normalization. In: Proc. SIGIR 1996, pp. 21–29 (1996)
15. http://www.grouplens.org/, http://www.movielens.org/

# Fast Similarity Computation
# in Factorized Tensors

Michael E. Houle[1], Hisashi Kashima[2,3], and Michael Nett[1,2]

[1] National Institute of Informatics, Tokyo 101-8430, Japan
{meh,nett}@nii.ac.jp
[2] University of Tokyo, Tokyo 113-8656, Japan
kashima@mist.i.u-tokyo.ac.jp
[3] Basic Research Programs PRESTO,
Synthesis of Knowledge for Information Oriented Society

**Abstract.** Low-rank factorizations of higher-order tensors have become an invaluable tool for researchers from many scientific disciplines. Tensor factorizations have been successfully applied for moderately sized multimodal data sets involving a small number of modes. However, a significant hindrance to the full realization of the potential of tensor methods is a lack of scalability on the client side: even when low-rank representations are provided by an external agent possessing the necessary computational resources, client applications are quickly rendered infeasible by the space requirements for explicitly storing a (dense) low-rank representation of the input tensor. We consider the problem of efficiently computing common similarity measures between entities expressed by fibers (vectors) or slices (matrices) within a given factorized tensor. We show that after appropriate preprocessing, inner products can be efficiently computed independently of the dimensions of the input tensor.

**Keywords:** similarity computation, inner products, tensor factorization.

## 1 Introduction

Within many emerging areas of computing, such as data mining, recommendation systems, security, and multimedia, applications of similarity search naturally arises in the context of such fundamental tasks as clustering, classification, matching and detection. In each case, the data model must reflect the underlying temporal, spatial, social, or other relationships that may be important for the application at hand. In many instances, the complexity of the data relationships precludes a simple attribute-vector representation. For example, a recommender system for movies might involve several modes of information, each with their own sets of attributes, in order to model the interactions among such entities as users, movies, ratings, dates and times, and so on. Due to the complexity inherent in managing multimodal data, effective and scalable strategies for similarity search are crucial to the overall performance of the system.

G. Navarro and V. Pestov (Eds.): SISAP 2012, LNCS 7404, pp. 226–239, 2012.

Multimodal data can be naturally represented in the form of a tensor (also known as a multiway array), a higher-dimensional extension of the matrix representation. Tensor-based data modeling is particularly appealing whenever the data dynamics can be captured by truncated (low-rank) representations, in terms of a small number of latent variables. Such representations, obtained by means of different tensor factorization models, have already become commonplace in areas including recommender systems [18], trend analysis [9], signal processing [8], analytical chemistry [2], psychometrics [17], graph and network analysis [19,11] and computer vision [30]. The popularity of tensor factorization in practice has led to several surveys and books on the topic [20,27,28]. The development of tensor factorization models can be dated back to Hitchcock in 1927 [16]; since then, many models have been proposed, with significant contributions by Carroll and Change [7], Harshman [15], Tucker [29] and Rendle and Schmidt-Thieme [26].

As tensors themselves constitute a structural generalization of matrices to higher orders, we may consider matrix factorization models as the conceptual ancestor of their tensor-based counterparts. The undoubtedly most popular matrix factorization model is the *singular value decomposition* (SVD), several implementations of which have appeared in recent years [23]. Applications of singular value decomposition have become commonplace in such areas as pattern recognition, security, recommender systems, text mining, and bioinformatics.

Recent attention has been given to the issue of efficiency and scalability of the tensor factorization process itself [1,2], including memory efficiency [22] and the efficiency of tensor arithmetic [22,3]. Tensor factorizations have been successfully applied for moderately sized multimodal data sets involving a small number of modes. However, relevant data sets in, for example, real-world recommender system applications [6,5] are typically much larger, in both the number of observations and the modality. One way in which practitioners solve the inherent problem of the space requirements associated with higher-order tensors is by considering only a core group of relations so as to improve the tractability of the tensor factorization processes. A core group would allow, for example, only data from objects that have a sufficiently high number of interactions within a system (such as among users, items, and a restricted set of tags).

Even though for applications such as recommender systems the observations typically generate data that admit a sparse representation, the intermediate representations used in processing the data are usually dense. In such situations, even if a low-rank representation of a large input relation were to be provided by an external agent possessing the necessary computational resources, client applications would nevertheless be rendered infeasible by the space requirements for explicitly maintaining the dense tensors formed at intermediate steps of the computation. This problem is usually addressed only by maintaining a low-rank factorization of the input tensor itself. Depending on the model rank, this would allow significant reductions in the storage requirements, but also incur a significant additional computational cost each time an individual tensor element is reconstituted from the tensor factors.

In this paper we consider the problem of efficiently computing similarity scores for multimodal data represented as a tensor factorization, without the need to explicitly reconstitute the individual elements of the input tensor. Here, we assume that the similarity measures can be expressed in terms of inner products between fibers (vectors) and slices (matrices) of a tensor that is maintained in form of a factorization of low rank. Inner products constitute the building blocks for many popular measures of similarity, such as the Euclidean distance or the cosine similarity. For several popular tensor factorization models, we show how data obtained in an additional preprocessing step can be utilized to efficiently evaluate inner products, with a computational complexity that depends only on the choice of model parameters, rather than on the dimensions of the input tensor. By avoiding the explicit reconstruction of the original tensor values, significant savings in both computational time and storage can be achieved.

The remainder of the paper is structured as follows. Section 2 briefly reviews the notation and concepts needed for the discussion of tensors and inner products. In Section 3 we summarize three popular tensor factorization models, and in Section 4 we propose our computation scheme for efficiently computing the similarity between tensor substructures. An empirical evaluation of our method appears in Section 5. The paper concludes in Section 6 with a discussion of potential directions for future research.

## 2     Preliminaries

This section introduces the basic concepts and terminology used in this paper. Where convenient, we adhere to the notation used in [20].

### 2.1     Vectors, Matrices and Tensors

An *order-p* tensor over a field $R$ is the product of $p$ individual $R$-vector spaces; accordingly, an order-1 (first-order) tensor is a vector, and an order-2 (second-order) tensor is a matrix. We denote vectors by bold lower-case letters, matrices by bold capitals and order-$p$ tensors ($p > 2$) using the Fraktur typeface. For any matrix $\mathbf{U}$, the vector $\mathbf{u}_k$ refers to its $k$-th column vector, and $u_{ij}$ refers to the matrix coefficient located in row $i$ and column $j$. Given an order-$p$ tensor, fixing all but the $k$-th index ($1 \leq k \leq p$) produces a *mode-k fiber*, a vector of length equal to the $k$-th mode of the tensor. Given a tensor $\mathfrak{X} = [x_{i_1 \cdots i_p}]_{n_1 \times \cdots \times n_p}$, we will denote individual mode-$k$ fibers using notation of the form $\mathbf{x}_{i_1 \cdots i_{k-1} \square i_{k+1} \cdots i_p}$. Analogously, one may denote higher-order subtensors by introducing additional wild-card symbols ($\square$) into the list of indices.

Given vectors $\mathbf{u}_i \in \mathbb{R}^{n_i}$ for $1 \leq i \leq p$, the *outer product* $\mathbf{u}_1 \circ \cdots \circ \mathbf{u}_p$ (see Fig. 1) is an order-$p$ tensor $\mathfrak{U} \in \mathbb{R}^{n_1 \times \cdots \times n_p}$ whose elements are defined by

$$u_{i_1 \cdots i_p} = \prod_{j=1}^{p} u_{j i_j}, \text{ where } 1 \leq i_j \leq n_j \text{ for } 1 \leq j \leq p. \tag{1}$$

$$\circ \begin{bmatrix} ③ & 1 & 0 & 2 \end{bmatrix}$$

$$\begin{bmatrix} 1 \\ ② \\ 0 \\ 1 \end{bmatrix} \qquad = \qquad \begin{bmatrix} 3 & 1 & 0 & 2 \\ ⑥ & 2 & 0 & 4 \\ 0 & 0 & 0 & 0 \\ 3 & 1 & 0 & 2 \end{bmatrix}$$

**Fig. 1.** The outer product of two vectors produces a second-order tensor (a matrix). The matrix element at location $(i, j)$ is the product of the $i$-th element of the first vector with the $j$-th element of the second vector.

Given a pair of vectors $\mathbf{u}$ and $\mathbf{v}$, one may simply write $\mathbf{u} \circ \mathbf{v} = \mathbf{u}\mathbf{v}^\top$.

Any order-$p$ tensor is said to be of rank one if and only if it may be expressed as an outer product of the form $\mathbf{u}_1 \circ \cdots \circ \mathbf{u}_p$. More generally, the *tensor rank* of an arbitrary order-$p$ tensor $\mathfrak{X}$, $rk(\mathfrak{X})$, is defined as the minimum number of rank-1 order-$p$ tensors, such that their (element-wise) sum equals $\mathfrak{X}$. Note that in the case $p = 2$, this definition of rank conforms to the usual notion of matrix rank. Unfortunately, for $p > 2$, determining the tensor rank turns out to be *NP*-hard [14].

## 2.2 Inner Products and Norms

Let $\mathbf{u}_1, \ldots, \mathbf{u}_p$ be vectors of length $n$ over some field. The *general inner product* $\Phi(\mathbf{u}_1, \ldots, \mathbf{u}_p)$ is determined by

$$\Phi(\mathbf{u}_1, \ldots, \mathbf{u}_p) = \sum_{i=1}^n \prod_{j=1}^p u_{ji} \tag{2}$$

As with the outer product, for the special case when $p = 2$, one may abbreviate $\Phi(\mathbf{u}, \mathbf{v})$ by $\mathbf{u}^\top \mathbf{v}$ whenever convenient. Inner products are particularly interesting for similarity search applications, because they serve as the building blocks from which one may construct popular distance measures, such as the Euclidean distance

$$d_2(\mathbf{u}, \mathbf{v}) = \sqrt{\Phi(\mathbf{u}, \mathbf{u}) + \Phi(\mathbf{v}, \mathbf{v}) - 2 \cdot \Phi(\mathbf{u}, \mathbf{v})} \tag{3}$$

or the cosine similarity

$$s(\mathbf{u}, \mathbf{v}) = (\Phi(\mathbf{u}, \mathbf{u}) \cdot \Phi(\mathbf{v}, \mathbf{v}))^{-\frac{1}{2}} \cdot \Phi(\mathbf{u}, \mathbf{v}) \tag{4}$$

The *Frobenius norm* of a tensor $\mathfrak{X} \in \mathbb{R}^{n_1 \times \cdots \times n_p}$ may be conveniently defined in terms of its mode-$k$ fibers, for any desirable choice of $1 \le k \le p$:

$$\|\mathfrak{X}\|_F = \sqrt{\sum_{i_1=1}^{n_1} \cdots \sum_{i_{k-1}=1}^{n_{k-1}} \sum_{i_{k+1}=1}^{n_{k+1}} \cdots \sum_{i_p=1}^{n_p} \Phi(\mathbf{x}_{i_1 \cdots i_{k-1} \Box i_{k+1} \cdots i_p}, \mathbf{x}_{i_1 \cdots i_{k-1} \Box i_{k+1} \cdots i_p})} \tag{5}$$

$$\hat{\mathfrak{x}} = \begin{bmatrix} 4 & 6 & 11 \\ 2 & 2 & 5 \\ 3 & ③ & 4 \end{bmatrix} = \begin{bmatrix} 1 & 0 & 2 \\ 1 & 0 & 2 \\ 0 & ⓪ & 0 \end{bmatrix} + \begin{bmatrix} 3 & 6 & 9 \\ 1 & 2 & 3 \\ 1 & ② & 3 \end{bmatrix} + \begin{bmatrix} 0 & 0 & 0 \\ 0 & 0 & 0 \\ 2 & ① & 1 \end{bmatrix}$$

$$\circ \begin{bmatrix} 1 & ⓪ & 2 \end{bmatrix} \qquad \circ \begin{bmatrix} 1 & ② & 3 \end{bmatrix} \qquad \circ \begin{bmatrix} 2 & ① & 1 \end{bmatrix}$$

$$= \begin{bmatrix} 1 \\ 1 \\ ⓪ \end{bmatrix} \qquad + \begin{bmatrix} 3 \\ 1 \\ ① \end{bmatrix} \qquad + \begin{bmatrix} 0 \\ 0 \\ ① \end{bmatrix}$$

**Fig. 2.** Depiction of a rank-3 CP decomposition of a second-order tensor. For convenience, we assume that the weights $w_i$ are equal to 1. The circled entries illustrate the decomposition of an individual tensor element.

## 3   Factorization Models

In this section we present three popular tensor factorization models: the canonical polyadic (CP) decomposition, the TUCKER factorization and the pairwise interaction factorization (PITF) model. We provide short formal descriptions of each model, briefly discuss their particular properties, and indicate their publicly available implementations. We also discuss how individual tensor elements can be reconstituted from a given factorization.

### 3.1   Canonical Polyadic Decomposition

The *canonical polyadic* (CP) decomposition model (also known as *parallel factorization*) dates back to Hitchcock [16]. While it has initially been applied in psychometrics and analytical chemistry [2,7,15], nowadays it is also an invaluable tool in bioinformatics [24], signal processing [10] and web analysis [21]. A CP factorization describes an arbitrary order-$p$ tensor $\mathfrak{X}$ as the conical combination of rank-1 tensors $\mathfrak{X}^{[1]}, \dots, \mathfrak{X}^{[m]}$. As rank-1 tensors, the individual $\mathfrak{X}^{[i]}$ may be expressed in terms of vector products (see Fig. 2):

$$\mathfrak{X} = \sum_{i=1}^{m} w_i \mathfrak{X}^{[i]} = \sum_{i=1}^{m} w_i \mathbf{u}_1^{[i]} \circ \cdots \circ \mathbf{u}_p^{[i]} \tag{6}$$

The factorization is parameterized by the *model rank* $m$, which is chosen by the user. By definition it is not possible to decompose a tensor $\mathfrak{X}$ using only $m < rk(\mathfrak{X})$ terms; it is, however, possible to efficiently fit a *rank deficient* model $\hat{\mathfrak{X}}$ with respect to $\|\hat{\mathfrak{X}} - \mathfrak{X}\|_F$, the Frobenius norm of the residual. Note that this is equivalent to minimizing the element-wise mean squared error between $\mathfrak{X}$ and $\hat{\mathfrak{X}}$. In the particular case of $p = 2$, this is equivalent to a truncated singular value decomposition [12]. A generic implementation supporting both dense and sparse higher-order tensors is available as part of the Tensor Toolbox for MATLAB [4].

$$\mathfrak{C} = \begin{bmatrix} 1 & 1 \\ 1 & 0 \end{bmatrix}, \quad \mathbf{U}^{[1]} = \begin{bmatrix} 1 & 2 \\ 0 & 1 \\ 2 & 1 \end{bmatrix}, \quad \mathbf{U}^{[2]} = \begin{bmatrix} 1 & 2 \\ 0 & 2 \\ 0 & 1 \end{bmatrix}, \quad \hat{\mathfrak{X}} = \begin{bmatrix} 5 & 2 & 1 \\ 1 & 0 & 0 \\ 7 & 4 & 2 \end{bmatrix}$$

$$\langle \hat{x}_{31} \rangle = c_{11} u_{31}^{[1]} u_{11}^{[2]} + c_{12} u_{31}^{[1]} u_{12}^{[2]} + c_{21} u_{32}^{[1]} u_{11}^{[2]} + c_{22} u_{32}^{[1]} u_{12}^{[2]}$$

**Fig. 3.** A Tucker decomposition of the second order tensor $\hat{\mathfrak{X}} \in \mathbb{R}^{3 \times 3}$. The enclosing boxes indicate the decomposition of the tensor element $\hat{x}_{31}$. Unlike with the CP decomposition, all factors of a Tucker decomposition interact with each other.

According to Equation 6, in any such rank-$m$ decomposition of an order-$p$ tensor $\hat{\mathfrak{X}}$, one may reconstitute an individual tensor element $\hat{x}_{i_1 \cdots i_p}$ in time $\Theta(pm)$ using the formula

$$\hat{x}_{i_1 \cdots i_p} = \sum_{j=1}^{m} w_j \prod_{k=1}^{p} u_{k i_k}^{[j]}. \tag{7}$$

### 3.2 Tucker Decomposition

The *Tucker factorization model* (Tucker) was proposed by Ledyard Tucker as a form of higher-order singular value decomposition [29]. In his model, an order-$p$ tensor in $\mathfrak{X} \in \mathbb{R}^{n_1 \times \cdots \times n_p}$ is expressed as the 'mode-wise' product of an order-$p$ *core tensor* $\mathfrak{C}$ with $p$ individual factor matrices (see Fig. 3):

$$\mathfrak{X} = \mathfrak{C} \times_1 \mathbf{U}^{[1]} \times_2 \cdots \times_p \mathbf{U}^{[p]} = \sum_{i_1=1}^{n_1} \cdots \sum_{i_p=1}^{n_p} c_{i_1 \cdots i_p} \mathbf{u}_{i_1}^{[1]} \circ \cdots \circ \mathbf{u}_{i_p}^{[p]}, \tag{8}$$

where $\mathfrak{C}$ has the same dimensions as $\mathfrak{X}$, and the factor matrix $\mathbf{U}^{[i]}$ has size $n_i \times n_i$ for $1 \leq i \leq p$. As with the (second-order) singular value decomposition, the factor matrices are orthogonal. As with the CP model (see Section 3.1), it is also possible to fit a truncated model $\hat{\mathfrak{X}}$ using a user-provided core tensor size $m_1 \times \cdots \times m_p$ satisfying $m_i < n_i$ for at least one instance of $1 \leq i \leq p$. We refer to the model rank of a Tucker decomposition by the size of its core tensor. An implementation of the Tucker model is available as part of the MatLab Tensor Toolbox [4].

From Equation 8, we may reconstitute any tensor element from a given rank-$(m_1, \ldots, m_p)$ factorization of $\hat{\mathfrak{X}}$ using

$$\hat{x}_{i_1 \cdots i_p} = \sum_{j_1=1}^{m_1} \cdots \sum_{j_p=1}^{m_p} c_{j_1 \cdots j_p} u_{j_1 i_1}^{[1]} \cdots u_{j_p i_p}^{[p]} \tag{9}$$

$$\hat{\mathfrak{x}} = \begin{bmatrix} 1 & 3 & 0 \\ 2 & 7 & 2 \\ \boxed{1} & 3 & 0 \end{bmatrix}, \quad \mathbf{U}^{[1,2]} = \begin{bmatrix} 1 & 2 & \boxed{1} \\ 0 & 1 & \boxed{0} \end{bmatrix}, \quad \mathbf{U}^{[2,1]} = \begin{bmatrix} 1 & \boxed{3} & 0 \\ 0 & \boxed{0} & 0 \end{bmatrix}$$

**Fig. 4.** A rank-2 PITF model of the second-order tensor $\hat{\mathfrak{x}} \in \mathbb{R}^{3\times3}$. The boxed elements indicate the decomposition of an individual tensor element. Note that, for the special case $p = 2$, the tensor $\hat{\mathfrak{x}}$ is merely the product of the two factor matrices.

This process requires a number of steps proportional to the volume of the core tensor: $m_1 \cdots m_p$. When assuming a hypercubic core tensor $\mathfrak{C} \in \mathbb{R}^{m\times\cdots\times m}$, the total time required is in $\Theta(pm^p)$.

The TUCKER model may be seen as a generalization of the CP model (see Section 3.1) in that it allows arbitrary interactions between individual columns of the factor matrices. While this increases the expressiveness of the model, it also significantly increases the complexity associated with accessing its elements (as can be seen from Equation 9).

### 3.3   Pairwise Interaction Model

The *pairwise interaction* (PITF) model originates from a personalized tag recommendation system in which the relationships among users, items and tags are modeled as a third-order tensor [26]. PITF may be seen as a special case of the canonical polyadic model for third-order tensors, with the constraint that for any given additive term, factor matrices from at most two different modes may interact. In the original model, a tensor element $\hat{y}_{uit}$ corresponding to a user $u$, an item $i$ and a tag $t$ is produced by (see [26])

$$\hat{y}_{uit} = \sum_f \hat{u}_{uf}^T \hat{t}_{tf}^U + \sum_f \hat{i}_{if}^T \hat{t}_{tf}^I + \sum_f \hat{u}_{uf}^I \hat{i}_{if}^U. \tag{10}$$

The authors also provide BPR-OPT, an implementation of a sampling-based gradient descent algorithm for efficiently learning the factor matrices [25].

We generalize this model to higher-order settings ($p > 3$) by computing $\mathbf{U}^{[i,j]} \in \mathbb{R}^{n_i \times m}$ for each $i, j \in \{1, \ldots, p\}$ with $i \neq j$ (see Fig. 4). Extending the model to account for all possible pairs of interaction yields a rather unwieldy expression for $\hat{\mathfrak{x}}$. Fortunately, the formula according to which individual tensor elements are computed is quite simple:

$$\hat{x}_{i_1 \cdots i_p} = \sum_{k=1}^m \sum_{1 \leq a < b \leq p} u_{ki_a}^{[a,b]} u_{kj_b}^{[b,a]} \tag{11}$$

Note that the computation time required to reconstitute an element is proportional to $m\binom{p}{2}$. Conveniently, due to the restricted interaction between factor matrices, the computation of the 'sample gradient' in BPR-OPT can be trivially extended to our higher-order interpretation of the model.

| Complexity | Original | TUCKER | CP | PITF |
|---|---|---|---|---|
| Size | $\prod_i n_i$ | $\prod_i m_i + \sum_i m_i n_i$ | $m\sum_i n_i$ | $mp\sum_j n_i$ |
| Access | 1 | $p\prod_i m_i$ | $mp$ | $mp^2$ |
| Preprocessing | 1 | $\sum_i m_i^2$ | $m^2 p$ | $m^2 p^2$ |
| Naïve Inner Product | $n_k$ | $n_k p\prod_i m_i$ | $n_k mp$ | $n_k mp^2$ |
| Fast Inner Product | — | $pm_k\prod_i m_i$ | $m^2 p$ | $m^2 p^4$ |

**Fig. 5.** Complexities of the different factorization models, as compared to operations on the original unfactorized tensor. *Size* refers to the amount of memory required to store the factorization. *Access* is the cost of computing a single tensor element (see Section 4). The *Preprocessing* row refers to the additional storage requirements imposed by our preprocessing step. *Fast* and *Naïve Inner Product* refer to the complexity of computing inner products between fibers (vectors) with and without using information obtained in our proposed preprocessing step.

## 4    Fast Similarity Computation

The previous section introduced popular factorization models for higher-order tensors. As previously stated in [26], the model classes of the individual models constitute a hierarchy, with TUCKER being the most general model and CP being a special case of TUCKER. While the PITF model in its previously proposed (third-order) interpretation is a special case of the canonical polyadic model, its higher-order interpretation does not fit into this hierarchy.

Each of these three models comes with an intrinsic complexity that can be regarded as a baseline against which can be measured the complexity of tensor operations involved in similarity search, in particular the costs associated with reconstituting a single tensor element (see Fig. 5). This gives rise to an interesting question: need the computational cost of evaluating the similarity of two substructures (such as fibers or slices) depend strongly on the number of tensor elements accessed? In the remainder of this section we show how one can efficiently compute such similarity values with cost independent of the size of the input tensor.

### 4.1    Fast Inner Product Computation

Whenever limited resources allow the storage of only a factorization of a tensor, rather than an explicit tensor representation, any access to a tensor element comes at a cost associated with that factorization model. The naïve approach to compute the inner product of (say) two mode-$k$ fibers is to first retrieve all $2 \cdot n_k$ elements within the fibers, and then to compute the product value. However, the following preprocessing schemes allow a significant reduction in computational cost to be traded against a small increase in storage requirements.

- Given a rank-$m$ CP factorization of $\hat{\mathfrak{X}}$ (see Equation 6), evaluate and store the inner products $\Phi(\mathbf{u}_k^{[i]}, \mathbf{u}_k^{[j]})$ of pairs of columns of the factor matrix associated with mode $1 \leq k \leq p$, where $1 \leq i \leq j \leq m$. Fig. 6 shows how the

$$\Phi_{CP}(\hat{x}_{\square i_2 \cdots i_p}, \hat{x}_{\square j_2 \cdots j_p}) = \sum_{k=1}^{n_1} \hat{x}_{ki_2 \cdots i_p} \hat{x}_{kj_2 \cdots j_p}$$

$$= \sum_{k=1}^{n_1} \left( \sum_{a=1}^{m} w_a u_{1k}^{[a]} u_{2i_2}^{[a]} \cdots u_{pi_p}^{[a]} \right) \left( \sum_{b=1}^{m} w_b u_{1k}^{[b]} u_{2j_2}^{[b]} \cdots u_{pj_p}^{[b]} \right)$$

$$= \sum_{a,b=1}^{m} w_a w_b u_{2i_2}^{[a]} \cdots u_{pi_p}^{[a]} u_{2j_2}^{[b]} \cdots u_{pj_p}^{[b]} \sum_{k=1}^{n_1} u_{1k}^{[a]} u_{1k}^{[b]}$$

$$= \sum_{a,b=1}^{m} w_a w_b u_{2i_2}^{[a]} \cdots u_{pi_p}^{[a]} u_{2j_2}^{[b]} \cdots u_{pj_p}^{[b]} \Phi(u_1^{[a]}, u_1^{[b]})$$

**Fig. 6.** Computation of the inner product of mode-1 fibers in a rank-$m$ CP factorization. The derivation shows how one can evaluate inner products in $\Theta(m^2 p)$ time (see Fig. 5). This method generalizes to fibers of arbitrary modes simply by exchanging the roles of the indices in the summation.

information generated during preprocessing can be used to accelerate the computation of similarities.

- Similarly, for a given rank-$(m_1, \ldots, m_p)$ TUCKER factorization, compute and store $\Phi(u_i^{[k]}, u_j^{[k]})$ for each mode $k \in \{1, \ldots, p\}$ and $1 \leq i \leq j \leq m_k$. The derivations in Fig. 7 show how to use the information gained during preprocessing to speed up the computation of inner products.
- Provided a generalized PITF factorization, create look-up tables for inner products of the form $\Phi(u_a^{[i,j]})$ and $\Phi(u_a^{[i,j]}, u_b^{[i,k]})$ for any choice of $a, b \in \{1, \ldots, m\}$ and $i, j, k \in \{1, \ldots, p\}$. Fig. 8 demonstrates how this additional step can ease the evaluation of inner products of fibers.

The equations in Figs. 6–8 show that the dependency on the shape of the input tensor can be moved into a model-specific preprocessing step. Hence, by paying the relatively small cost associated with the proposed precomputation schemes, the cost of evaluating inner products of fibers becomes independent of the dimensions of the input tensor. Furthermore, these results may be generalized to higher-order substructures within the factorized tensor, such as (for example) second-order slices (see Section 4.2). In addition, when assuming $n_i = n_j$ for some $i \neq j$, we can also evaluate inner products of a mode-$i$ and mode-$j$ fiber with some additional preprocessing. In fact, if we have prior knowledge of those modes or combination of modes for which we wish to compute similarity scores, we can reduce even the cost of the preprocessing steps by computing only the relevant entries of the look-up tables.

## 4.2    Inner Products of Higher-Order Substructures

As described in Section 2, popular similarity measures such as the Euclidean distance or the cosine similarity between two mode-$k$ fibers $\mathbf{u}$ and $\mathbf{v}$ can be

$$\Phi_{Tucker}(\hat{x}_{\Box i_2 \cdots i_p}, \hat{x}_{\Box j_2 \cdots j_p})$$

$$= \sum_{k=1}^{n_1} \hat{x}_{k i_2 \cdots i_p} \hat{x}_{k j_2 \cdots j_p}$$

$$= \sum_{k=1}^{n_1} \sum_{k_1=1}^{m_1} \cdots \sum_{k_p=1}^{m_p} \hat{c}_{k_1 \cdots k_p} u_{k k_1}^{[1]} u_{i_2 k_2}^{[2]} \cdots u_{i_p k_p}^{[p]} \sum_{h_1=1}^{m_1} \cdots \sum_{h_p=1}^{m_p} \hat{c}_{h_1 \cdots h_p} u_{k h_1}^{[1]} u_{j_2 h_2}^{[2]} \cdots u_{j_p h_p}^{[p]}$$

$$= \sum_{k_1,h_1=1}^{m_1} \sum_{k=1}^{n_1} u_{k k_1}^{[1]} u_{k h_1}^{[1]} \sum_{k_2,h_2=1}^{m_2} \cdots \sum_{k_p,h_p=1}^{m_p} \hat{c}_{k_1 \cdots k_p} u_{i_2 k_2}^{[2]} \cdots u_{i_p k_p}^{[p]} \hat{c}_{h_1 \cdots h_p} u_{j_2 h_2}^{[2]} \cdots u_{j_p h_p}^{[p]}$$

$$= \sum_{k_1,h_1=1}^{m_1} \Phi(u_{k_1}^{[1]}, u_{h_1}^{[1]}) \sum_{k_2=1}^{m_2} \cdots \sum_{k_p=1}^{m_p} \hat{c}_{k_1 \cdots k_p} u_{i_2 k_2}^{[2]} \cdots u_{i_p k_p}^{[p]} \sum_{h_2=1}^{m_2} \cdots \sum_{h_p=1}^{m_p} \hat{c}_{h_1 \cdots h_p} u_{j_2 h_2}^{[2]} \cdots u_{j_p h_p}^{[p]}$$

**Fig. 7.** Computation of inner products of mode-1 fibers in a TUCKER factorization. The derivation shows how we can factor out the dependence on the shape of the input tensor with minimal additional memory requirements. We can perform this operation for a mode-$k$ fiber by interchanging the roles of the indices in the equations.

expressed as functions in $\Phi(\mathbf{u}, \mathbf{u})$, $\Phi(\mathbf{v}, \mathbf{v})$ and $\Phi(\mathbf{u}, \mathbf{v})$. In addition, we can generalize our results from the previous section to substructures of higher order. For example, the following equation demonstrates how inner products of mode-$(1, 2)$ slices (matrices) in a canonical rank-$m$ factorization can be efficiently computed.

$$\Phi(\hat{x}_{\Box\Box i_3 \cdots i_p}, \hat{x}_{\Box\Box j_3 \cdots j_p})$$

$$= \sum_{k_1=1}^{n_1} \sum_{k_2=1}^{n_2} \hat{x}_{k_1 k_2 i_3 \cdots i_p} \hat{x}_{k_1 k_2 j_3 \cdots j_p}$$

$$= \sum_{a,b=1}^{m} \Phi(\mathbf{u}_1^{[a]}) \Phi(\mathbf{u}_2^{[a]}) \Phi(\mathbf{u}_1^{[b]}) \Phi(\mathbf{u}_2^{[b]}) \prod_{f,g=3}^{p} u_{f i_f}^{[a]} u_{g j_g}^{[b]}$$

This scheme can be employed to compute the Frobenius norm of two different slices of a tensor, using the formulation of Equation 5. Note that we obtain terms of the form $\Phi(\mathbf{u})\Phi(\mathbf{v})$ rather than $\Phi(\mathbf{u}, \mathbf{v})$, since the two modes of the slice are independent. This process is equivalent to that of unfolding one mode of the tensor, such that the involved slices of size $n_i \times n_j$ become vectors of length $n_i n_j$, and subsequently applying the previously-discussed techniques to compute the Euclidean distance of the unfolded vectors. Again, the cost of that operation depends on model parameters, rather than the number of coefficients in the frontal slices. We must also account for the cost of computing the values of $\Phi(\mathbf{u})$; for the CP and TUCKER models, this introduces an additional precomputation cost of $\Theta(pm)$.

$$\Phi_{PITF}(\hat{x}_{\square i_2 \cdots i_p}, \hat{x}_{\square j_2 \cdots j_p})$$

$$= \sum_{k=1}^{n_1} \left( \sum_{f=1}^{m} \sum_{1 \le a < b \le p} u_{f i_a}^{[a,b]} u_{f i_b}^{[b,a]} \right) \left( \sum_{g=1}^{m} \sum_{1 \le c < d \le p} u_{g j_c}^{[c,d]} u_{g j_d}^{[d,c]} \right)$$

$$= \sum_{k=1}^{n_1} \sum_{f,g=1}^{m} \left( \sum_{\substack{1 < b \le p \\ 1 < d \le p}} u_{fk}^{[1,b]} u_{f i_b}^{[b,1]} u_{gk}^{[1,d]} u_{g j_d}^{[d,1]} + \sum_{\substack{1 < b \le p \\ 1 < c < d \le p}} u_{fk}^{[1,b]} u_{f i_b}^{[b,1]} u_{g j_c}^{[c,d]} u_{g j_d}^{[d,c]} \right.$$

$$\left. + \sum_{\substack{1 < a < b \le p \\ 1 < d \le p}} u_{f i_a}^{[a,b]} u_{f i_b}^{[b,a]} u_{g j_d}^{[d,1]} \sum_{k=1}^{n_1} u_{gk}^{[1,d]} + \sum_{\substack{1 < a < b \le p \\ 1 < c < d \le p}} u_{f i_a}^{[a,b]} u_{f i_b}^{[b,a]} u_{g j_c}^{[c,d]} u_{g j_d}^{[d,c]} \right)$$

$$= \sum_{f,g=1}^{m} \left( \sum_{\substack{1 < b \le p \\ 1 < d \le p}} u_{f i_b}^{[b,1]} u_{g j_d}^{[d,1]} \phi(\mathbf{u}_f^{[1,b]}, \mathbf{u}_g^{[1,d]}) + \sum_{\substack{1 < b \le p \\ 1 < c < d \le p}} u_{f i_b}^{[b,1]} u_{g j_c}^{[c,d]} u_{g j_d}^{[d,c]} \phi(\mathbf{u}_f^{[1,b]}) \right.$$

$$\left. + \sum_{\substack{1 < a < b \le p \\ 1 < d \le p}} u_{f i_a}^{[a,b]} u_{f i_b}^{[b,a]} u_{g j_d}^{[d,1]} \phi(\mathbf{u}_g^{[1,d]}) + \sum_{\substack{1 < a < b \le p \\ 1 < c < d \le p}} u_{f i_a}^{[a,b]} u_{f i_b}^{[b,a]} u_{g j_c}^{[c,d]} u_{g j_d}^{[d,c]} \right)$$

**Fig. 8.** Fast evaluation of inner products of mode-1 fibers in the generalized PITF model. The equations show how we can move all dependencies on the dimensions of the input tensor into the precomputation step. Again, mode-$k$ fibers may be handled by substituting indices.

## 5   Evaluation

We implemented a C++ framework for efficiently computing different similarity measures on fibers and matrices, from TUCKER, CP, or PITF tensor factorizations using the precomputation schemes presented earlier. We generate actual factorizations using the MATLAB Tensor Toolbox (for TUCKER and CP), or a generalization of the learning algorithm given in [26] (for the higher-order PITF model). We measured and compared the time consumed when computing inner products between pairs of fibers taken from different modes of an input tensor. The differences in the complexities of the individual models across different choices of model rank are shown in Fig. 9.

The time complexity of our computation scheme depends on the size of the factorization, but not on the content of the factor matrices nor on the number of entries of the input tensor. However, producing the factorizations themselves takes considerable execution time. To reduce the time required for our experimentation, we therefore analyzed our findings for only a subset of the Movielens data set [13]. Here, we included up to 8 variables (such as users, movies, and

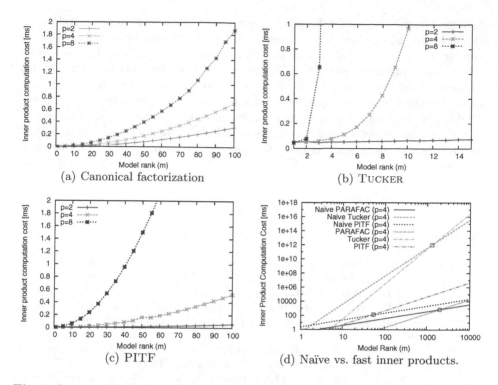

**Fig. 9.** Computational cost of evaluating inner products of fibers in different types of factorizations of order-$p$ tensors (Figs. 9(a)–9(c)), and a comparison of naïve vs. fast inner products for different choices of model rank $m$ (Fig. 9(d)).

tags), and restricted the modes of the tensor to have size 1000, by greedily selecting only the most dense regions of the data set.

In order to concisely present the experimental results, we chose the core tensor of the Tucker factorization to be a cube ($m = m_1 = \cdots = m_p$). For each choice of $p \in \{2, 4, 8\}$, we factorized our input tensor with respect to all three models using rank values ranging from 1 to 100. We limited the rank to a maximum of 15 for the Tucker model as its complexity makes it prohibitively expensive for larger model ranks (see Fig. 9(b)). Since the computation involves very little overhead, the time complexities shown in Fig. 5 are reflected in the experimental results in Figs. 9(a)–9(c). In addition, Fig. 9(d) shows the asymptotic trade-off between the dependence on the tensor dimensions and the model rank. Note that for most methods, the break-even points correspond to computation times that are already far in excess of tolerable limits.

## 6   Conclusion

In this paper, we proposed a computation scheme to accelerate computation of similarities between fibers and slices for low-rank representations of higher-order tensors. For the TUCKER, CP and the generalized PITF factorization models, we showed how the computation of inner products, Euclidean distances and cosine similarity values of fibers, as well as the Frobenius norm of the difference of two slices can be processed efficiently. Our approach provides exact similarity values between fibers and slices, while trading dependence on the size of the input tensor against a slightly larger dependence on the rank of the respective factorization model. Even though the most meaningful model rank still is an application-specific variable, our approach promises significant gains in scalability for many applications in areas such as machine learning, recommender systems or data warehousing. In fact, our experimental evaluation shows that our computation scheme outperforms a straightforward computation of similarity values, even for moderately large model ranks.

# References

1. Andersson, C.A., Henrion, R.: A general algorithm for obtaining simple structure of core arrays in $n$-way PCA with application to fluorometric data. Computational Statistics & Data Analysis 31(3), 255–278 (1999)
2. Appellof, C.J., Davidson, E.R.: Strategies for analyzing data from video fluorometric monitoring of liquid chromatographic effluents. Analytical Chemistry 53(13), 2053–2056 (1981)
3. Bader, B.W., Kolda, T.G.: Efficient MATLAB computations with sparse and factored tensors. SIAM Journal on Scientific Computing 30(1), 205–231 (2007)
4. Bader, B.W., Kolda, T.G.: MATLAB Tensor Toolbox Version 2.4 (March 2011), http://csmr.ca.sandia.gov/~tgkolda/TensorToolbox/
5. Bennett, J., Lanning, S., Netflix: The Netflix Prize. In: KDD Cup and Workshop in conjunction with KDD (2007)
6. Cantador, I., Brusilovsky, P., Kuflik, T.: 2nd workshop on information heterogeneity and fusion in recommender systems (HetRec 2011). In: Proceedings of the 5th ACM Conference on Recommender Systems, RecSys 2011, ACM (2011)
7. Carroll, J., Chang, J.-J.: Analysis of individual differences in multidimensional scaling via an $n$-way generalization of Eckart-Young decomposition. Psychometrika 35, 283–319 (1970)
8. Chen, B., Petropulu, A.P., Lathauwer, L.D.: Blind identification of convolutive MIMO systems with 3 sources and 2 sensors. EURASIP Journal on Advances in Signal Processing 5, 487–496 (2002)
9. Chi, Y., Tseng, B.L., Tatemura, J.: Eigen-trend: trend analysis in the blogosphere based on singular value decompositions. In: Proceedings of the 15th ACM International Conference on Information and Knowledge Management, CIKM, pp. 68–77 (2006)
10. Comon, P.: Independent component analysis, a new concept? Signal Processing 36(3), 287–314 (1994)
11. Dunlavy, D.M., Kolda, T.G., Acar, E.: Temporal link prediction using matrix and tensor factorizations. ACM Transactions on Knowledge Discovery from Data 5(2), Article 10, 27 pages (2011)
12. Eckart, C., Young, G.: The approximation of one matrix by another of lower rank. Psychometrika 1, 211–218 (1936)
13. GroupLens Research Group. GroupLens Research Data Sets (August 2011), http://www.grouplens.org/node/12

14. Håstad, J.: Tensor rank is NP-complete. Journal of Algorithms 11(4), 644–654 (1990)
15. Harshman, R.A.: Foundations of the PARAFAC procedure: models and conditions for an explanatory multi-modal factor analysis. UCLA Working Papers in Phonetics 16, 1–84 (1970)
16. Hitchcock, F.L.: The expression of a tensor or a polyadic as a sum of products. J. Math. 6, 164–189 (1927)
17. Kapteyn, A., Neudecker, H., Wansbeek, T.: An approach to $n$-mode components analysis. Psychometrika 51, 269–275 (1986)
18. Karatzoglou, A., Amatriain, X., Baltrunas, L., Oliver, N.: Multiverse recommendation: $n$-dimensional tensor factorization for context-aware collaborative filtering. In: Proceedings of the 4th ACM Conference on Recommender Systems, pp. 79–86 (2010)
19. Kolda, T., Bader, B.: The TOPHITS model for higher-order web link analysis. In: Proceedings of the SIAM Data Mining Conference Workshop on Link Analysis, Counterterrorism and Security (2006)
20. Kolda, T.G., Bader, B.W.: Tensor decompositions and applications. SIAM Review 51(3), 455–500 (2009)
21. Kolda, T.G., Bader, B.W., Kenny, J.P.: Higher-order web link analysis using multilinear algebra. In: ICDM 2005: Proceedings of the 5th IEEE International Conference on Data Mining, pp. 242–249 (November 2005)
22. Kolda, T.G., Sun, J.: Scalable tensor decompositions for multi-aspect data mining. In: ICDM 2008: Proceedings of the 8th IEEE International Conference on Data Mining, pp. 363–372 (December 2008)
23. Lin, Z.: Some software packages for partial SVD computation. Computing Research Repository (CoRR) abs/1108.1548 (2011)
24. Möcks, J.: Topographic components model for event-related potentials and some biophysical considerations. IEEE Transactions on Biomedical Engineering 35(6), 482–484 (1988)
25. Rendle, S.: Tag recommender (October 2011), http://cms.uni-konstanz.de/informatik/rendle/software/tag-recommender/
26. Rendle, S., Schmidt-Thieme, L.: Pairwise interaction tensor factorization for personalized tag recommendation. In: WSDM, pp. 81–90 (2010)
27. Ren, Henrion: N-way principal component analysis theory, algorithms and applications. Chemometrics and Intelligent Laboratory Systems 25(1), 1–23 (1994)
28. Smilde, A.K., Wang, Y., Kowalski, B.R.: Theory of medium-rank second-order calibration with restricted-Tucker models. Journal of Chemometrics 8(1), 21–36 (1994)
29. Tucker, L.: Some mathematical notes on three-mode factor analysis. Psychometrika 31, 279–311 (1966)
30. Vasilescu, M.A.O., Terzopoulos, D.: Multilinear Analysis of Image Ensembles: TensorFaces. In: Heyden, A., Sparr, G., Nielsen, M., Johansen, P. (eds.) ECCV 2002, Part I. LNCS, vol. 2350, pp. 447–460. Springer, Heidelberg (2002)

# SIR: The Smart Image Retrieval Engine*

Jakub Lokoč, Tomáš Grošup, and Tomáš Skopal

SIRET Research Group, Dept. of Software Engineering,
Faculty of Mathematics and Physics, Charles University in Prague
{lokoc,skopal}@ksi.mff.cuni.cz, tomasgrosup@gmail.com

**Abstract.** We present the Smart Image Retrieval meta-search engine
that allows content-based exploration of the results obtained from various sources (mostly based on keyword query). The online feature extraction architecture and exploration models utilizing single-/multi-query
approaches are the two key features of our demo application that shows
very promising results.

## 1 Introduction

In the *Smart Image Retrieval* (SIR) [3] demo application, we develop techniques
contributing to visual exploration of results obtained by keyword image search,
and their subsequent implementation within a prototypical image meta-search
web engine. In our approach, the crawling, storage and persistent indexing is
not necessary, because the content-based similarity search techniques are evaluated only on the top results returned by keyword-based search engines. This
meta-search approach has been already investigated in the computer vision domain (e.g. in [2] or [4]), where the top results are filtered using content-based
methods and employed in automatic learning of visual models. The SIR engine
focuses on the image exploration, where a keyword meta-query is passed to image search engines using full-text search, such as Google Images or Bing Images.
The obtained result (ranked images) is then processed to incorporate also a
content-based image retrieval (CBIR) semantics into the search process.

In order to re-rank the images based on visual content, a feature extraction
technique has to be used to obtain some visual information from the candidate
images. In case of a traditional CBIR engine, the feature extraction and content-based indexing would be performed on usually large dataset in the offline phase,
that is, before queries are issued by users. On the other hand, in the meta-search
engine we do not maintain any content-based index and so we need to extract
the content features from relatively small dataset during the query processing
(i.e., online). In our approach we consider the MPEG-7 visual descriptors [5] and
so-called image feature signatures [1]. Once a ranking of images is returned from
the other search engines given a keyword query, a feature extraction on the top
images is executed. This allows utilization of various kinds of similarity models,

---

* This research has been supported in part by Czech Science Foundation projects
P202/11/0968 and P202/12/P297.

that are not fixed to an extracted features or already built indexes. Because of the online nature the feature extraction needs to perform very fast, so that the user is not noticeably delayed. For example, a GPU parallelism is a suitable platform for such an instant online feature extraction.

In Figure 1 see the basic exploration layout used in the SIR engine for the presentation of the query result. The layout support various exploration capabilities, such as zooming in/out, panning, multi-query refinement, etc.

**Fig. 1.** Exploration layout used in the SIR engine for keyword query 'Jaguar'

# References

1. Beecks, C., Uysal, M.S., Seidl, T.: Signature quadratic form distance. In: Proc. ACM CIVR, pp. 438–445 (2010)
2. Fergus, R., Fei-Fei, L., Perona, P., Zisserman, A.: Learning object categories from google"s image search. In: IEEE Int. Conf. on Computer Vision, ICCV 2005, pp. 1816–1823. IEEE Computer Society, Washington, DC (2005)
3. Grošup, T., Lokoč, J., Skopal, T.: Smart Image Retrieval (SIR), SIRET Research Group (2012), http://www.siret.cz/sir
4. Li, L.-J., Fei-Fei, L.: Optimol: Automatic online picture collection via incremental model learning. Int. J. Comput. Vision 88(2), 147–168 (2010)
5. Manjunath, B.S., Salembier, P., Sikora, T. (eds.): Introduction to MPEG-7: Multimedia Content Description Interface. John Wiley & Sons, Inc. (2002)

# SimTandem: Similarity Search
# in Tandem Mass Spectra*

Jiří Novák, Jakub Galgonek, David Hoksza, and Tomáš Skopal

Siret Research Group,
Faculty of Mathematics and Physics, Charles University in Prague,
Malostranské nám. 25, 118 00 Prague, Czech Republic
novak@ksi.mff.cuni.cz

**Abstract.** SimTandem is a tool for fast identification of protein and peptide sequences from tandem mass spectra. The identification is based on similarity search of spectra captured by a tandem mass spectrometer in databases of theoretical mass spectra generated from databases of known protein sequences. Since the number of protein sequences in the databases grows rapidly and a sequential scan over the entire database of spectra is time-consuming, the non-metric access methods are employed as the database indexing techniques. SimTandem is based on a previously proposed method and is freely available at http://www.simtandem.org or http://www.siret.cz/simtandem.

**Keywords:** protein sequences identification, tandem mass spectrometry, similarity search, non-metric access methods, SimTandem.

## 1 Introduction

Proteins are the basis of all organisms securing almost every process on the cell level. The protein function is determined by its 3D structure while the structure is derived from the protein sequence. Tandem mass spectrometry (MS/MS) is a method for protein or peptide (a short piece of protein) sequences identification from an "in vitro" sample. A mass spectrometer captures a set of peptide mass spectra for a few proteins in the sample. Afterwards, the sequences must be identified from the spectra, e.g., by means of the similarity search in a database of theoretical spectra generated from a database of known protein sequences.

Although there are tools based on the similarity search in databases of protein sequences like MASCOT, SEQUEST or OMSSA [3], the designing of new algorithms for fast and accurate identification of sequences from the mass spectra is still desirable because there are many inaccuracies in the spectra and because the sizes of protein sequence databases grows rapidly.

* This work was supported by Czech Science Foundation (GAČR) projects 202/11/0968, 201/09/H057, by the Grant Agency of Charles University (GAUK) project Nr. 430711 and by the grant SVV-2012-265312.

G. Navarro and V. Pestov (Eds.): SISAP 2012, LNCS 7404, pp. 242–243, 2012.

One of the possibilities how to handle this growth is the application of various database indexing techniques. A few approaches were proposed where metric access methods were employed as the database indexing techniques and the cosine similarity (angle distance) was used as the mass spectra similarity function [4], [1]. A disadvantage is that the search of spectra with posttranslational modifications is not supported, even though the modifications are a common problem when peptide sequences are identified from the spectra.

## 2    Methods and Implementation

We have proposed an approach based on the non-metric access methods which enables fast and approximative identification of peptide sequences [3]. The parameterized Hausdorff distance is employed instead of the angle distance because the number of identified peptide sequences is higher and because the indexability by the non-metric access methods is better. Moreover, the approach supports the identification of spectra with posttranslational modifications.

Since a mass spectrometer captures a set of mass spectra for a few proteins in the sample, an extension of this approach was proposed for the identification of protein sequences from a set of mass spectra instead of the identification of peptide sequences from single spectra [2]. Since protein sequences contain more peptide sequences and the search in the non-metric index is followed by a sequential scan of protein sequence candidates, the number of identified peptide sequences is significantly increased.

SimTandem is an on-line tool for protein and peptide sequences identification from tandem mass spectra which implements the non-metric access methods as the database indexing techniques. The core of SimTandem is implemented in C++ and it employs the Siret Object Library (SOL) – a framework for efficient metric and non-metric similarity search which is currently being developed by Siret Research Group (SRG – http://www.siret.cz). SimTandem uses Intel's Threading Building Blocks (TBB) to support the parallel processing of large query sets of mass spectra. The web interface is implemented in Java based Google Web Toolkit (GWT). The communication between the web interface and the core is realized using Java RMI over IIOP and CORBA.

## References

1. Dutta, D., Chen, T.: Speeding up Tandem Mass Spectrometry Database Search: Metric Embeddings and Fast Near Neighbor Search. Bioinf. 23(5), 612–618 (2007)
2. Novák, J., Hoksza, D., Lokoč, J., Skopal, T.: On Optimizing the Non-metric Similarity Search in Tandem Mass Spectra by Clustering. In: Bleris, L., Măndoiu, I., Schwartz, R., Wang, J. (eds.) ISBRA 2012. LNCS, vol. 7292, pp. 189–200. Springer, Heidelberg (2012)
3. Novák, J., Skopal, T., Hoksza, D., Lokoč, J.: Non-metric Similarity Search of Tandem Mass Spectra Including Posttranslational Modifications. Journal of Discrete Algorithms 13, 19–31 (2012)
4. Ramakrishnan, S.R., et al.: A Fast Coarse Filtering Method for Peptide Identification by Mass Spectrometry. Bioinf. 22(12), 1524–1531 (2006)

# Author Index